Veterinary Mycology

Indranil Samanta

Veterinary Mycology

Springer

Indranil Samanta
Department of Veterinary Microbiology
West Bengal University of Animal & Fishery Sciences
Kolkata, West Bengal, India

ISBN 978-81-322-2279-8 ISBN 978-81-322-2280-4 (eBook)
DOI 10.1007/978-81-322-2280-4

Library of Congress Control Number: 2015930190

Springer New Delhi Heidelberg New York Dordrecht London

Printed on acid-free paper

Springer (India) Pvt. Ltd. is part of Springer Science+Business Media (www.springer.com)

To Baba, Ma, Jhuma, Subhranil for love and support
and
To my beloved students for inspiration

Foreword

Dr. Indranil Samanta is a young scientist who has gotten very important scientific achievements in his short carrier: Bachelor of Veterinary Sciences and Animal Husbandry, Master of Veterinary Sciences, Ph.D. in Veterinary Microbiology, and now he is Assistant Professor in the Microbiology Department of West Bengal University of Animal and Fishery Sciences, in India. He was also Assistant Professor, Division of Microbiology and Immunology, S. K. University of Agricultural Sciences and Technology of Kashmir, India, and Veterinary Officer of West Bengal Government. He received four awards and six research grants. His experience as publisher is wide, he belongs to the Editorial Board of six journals and acts as reviewer in other four journals, and he published 22 research articles in international journals. Last year he published a book entitled *Veterinary Bacteriology*.

Dr. Samanta kindly invited me to write the foreword of *Veterinary Mycology* and it is a great pleasure for me to present this book.

Medical and veterinary mycology have suffered very important transformations in the last three decades. The introduction of molecular biology techniques in taxonomy, epidemiology and diagnosis procedures, not based on cultures, was probably the most significant of them. The acquisition of more precise knowledge about pathogenesis of fungal infections was also very important in the management of these diseases. New antifungal drugs were incorporated to the therapeutic arsenal to fight against these infections, including new presentations of classic drugs as lipid formulations of amphotericin B and new compounds as echinocandins and second generation triazoles. These advances, as well as the increasing morbidity and mortality generated by the mycoses, attracted the attention of a great number of animal health professionals to veterinary mycology.

Those who are interested in this discipline will find in *Veterinary Mycology* an excellent guide book to increase their knowledge of fungal infections and their etiologic agents. This book is divided in three main parts: in the first one the history of veterinary mycology, general aspects of morphology, taxonomy and biology of fungi are considered. In the second part, the etiologic agents of superficial, deep, systemic and opportunistic mycoses are described with great detail. Biological aspects of these fungi, epidemiology of these infections, the immunity response of the hosts and the modern diagnosis techniques such as those searching for fungal antigens in organic fluid and those which applied molecular biology are extensively exposed. Clinical manifestations and therapy of the mycoses are presented in a more synthetic way, using tables containing comprehensive information. The third part is related to laboratory diagnosis including clinical samples collection and their processing for fungal isolation, special stains for fungal microscopic visualization and culture media composition. There are also special chapters about very infrequent fungal 'infection' and a glossary.

This book is written in concise and clear English, very easy to read. I think the readers will enjoy it very much.

Buenos Aires University School of Medicine
Buenos Aires, Argentina
Hospital F.J. Muñiz
Buenos Aires City, Argentina

Ricardo Negroni

Preface

The study of fungi began during ancient times. Even in early Sanskrit literature (*Atharva Veda*), scripts of Hippocrates and Lord Buddha and in the Holy Bible, the importance of mycology was mentioned for prevention of fungal diseases. Although it was little neglected both in medical and veterinary sciences, studying mycology is gaining importance in recent times due to emergence and re-emergence of fungal infection in human, animals and birds. Emergence of black yeasts in poultry, *Prototheca* in pets, *Laczia loboi* in human and marine animals, *Lagenidium* in pets, *Emmonsia* in human and pets as well as re-emergence of brooder pneumonia in poultry, candidiasis in human and animals, cryptococcosis in human and animals, and dermatophytosis in animals is noted in recent times. The fungal infection causes major economic loss in poultry and livestock related industry and it poses zoonotic threat especially to the pet owners. Advancement of knowledge helps in better understanding of the subject. Cumulation of the advancements along with the conventional knowhow in the area of veterinary mycology within the same cover was one of my best intentions. I have tried to restructure the classification, genome characteristics, pathogenesis, immunity, diagnosis and treatment of fungal diseases with the enlightened vision of molecular biology and discovery of new antifungals. I hope the information will be useful as reference text for undergraduate students, text for post-graduate students, veterinary practitioners and in allied sectors. Any correction, modification and suggestion for improvement are most welcome by the author.

During this auspicious occasion, I heartily acknowledge the mycology faculty members and scientists throughout the world who are stalwarts in this subject for evaluation of my chapters and their contribution of fungal photographs. I offer my sincere thanks to Professor Dr. Ricardo Negroni for his encouraging words. I also acknowledge friends and senior colleagues of my department and university for their valuable suggestions. All the schematic diagrams drawn by my wife Jhuma are duly accredited.

Kolkata, West Bengal, India
19 November 2014

Dr. Indranil Samanta

Contents

About the Author

Dr. Indranil Samanta obtained his Bachelor of Veterinary Sciences and Animal Husbandry degree (B.V.Sc. and A.H.) from West Bengal University of Animal and Fishery Sciences, Kolkata, India. He secured his Masters in Veterinary Sciences (M.V.Sc.) in Veterinary Bacteriology and Mycology from Indian Veterinary Research Institute, Bareilly, UP, India, and Doctor of Philosophy (Ph.D.) in Veterinary Microbiology from West Bengal University of Animal and Fishery Sciences, Kolkata, India. He works currently as Assistant Professor of Veterinary Microbiology in West Bengal University of Animal and Fishery Sciences, Kolkata, India. Previously he has also worked as Assistant Professor of Veterinary Microbiology in S. K. University of Agricultural Sciences and Technology-Kashmir, India, and as Veterinary Officer, Government of West Bengal, India. He is actively engaged in teaching of undergraduate, post-graduate and Ph.D. scholars of Veterinary Microbiology and research related with animal health and zoonotically important microbes. He has received six grants from national funding agencies and he has supervised three post-graduate scholars till date. He has published 70 research articles in reputed international and national journals along with review articles in international journals. His current total impact factor, h-index and total citations are 45, 10 and 280, respectively. He has published a textbook entitled *Veterinary Bacteriology* (ISBN13: 9789381450550, ISBN 10: 9381450552) from a reputed publisher. He is editorial board member and reviewer of international and national journals. He has delivered several talks in conferences, Government television and radio channels regarding his research and animal, poultry health related issues. He is the recipient of National Academy of Agricultural Sciences (NAAS) Associate, Government of India.

History 1

1.1 Medical Mycology

The term 'mycology' was coined by H.S. Berkley (1834) as a study of fungi. Medical or veterinary mycology is the study of medically or veterinary important fungi and fungal diseases in human and animals, respectively. The term 'mycosis' (mykes = mushroom) is used to describe the infection of human, animal, birds and plants which is caused by numerous pathogenic fungi. The 'mycotoxicosis' describes the diseased condition produced by the ingestion of mycotoxins (intoxication) present in the feed. When the fungi produce the pathogenesis due to in vivo toxin production after entry within the host, it is known as 'mycetism'.

In ancient Sanskrit writing of India (*Atharva Veda*, 2000–1000 BC), the first description of a fungal infection, i.e. mycetoma (*Pada Valmikan*), was documented. In seventeenth century, a German physician (Engelbert Kaempfer) working in India first reported clinical human cases of mycetoma which was followed by case reports of French missionaries in Pondicherry, India (ant-hill of worms, 1714). The fungal plant diseases (smut, rust) were noted in other *Veda* (1200 BC) also. In *Rigveda* crushing of mushroom (an edible fungus) with the feet was illustrated as a punishment. Hippocrates (460–377 BC) first documented oral pseudomembranous candidiasis and he described it with the name of 'aphthae albae' which was later supported by Galen (130–200 BC). Lord Buddha (400 BC) observed the fungal diseases of rice and sugarcane which can decrease their production. In the Bible (Amos 4, 19) instructions were provided to the priests for treatment of fungal infection such as dry rot.

The first description of dermatophytosis was recorded by Celsus, a Roman encyclopaedist who described a suppurative infection of the scalp ('porrigo' or 'kerion of Celsus') in *De Re Medicina* (30 AD). Throughout the middle ages several descriptions of dermatophytosis are produced where it was described as 'tinea' (Latin term). Micheli (1729) published *Nova genera Plantarum* (written in copper plate) in which he established several genera of fungi such as *Aspergillus*, *Mucor*, etc. However, the pathogenic potentiality of fungi in human or animals remained uncertain.

Robert Hook (1665) first illustrated the pathogenic role of rose rust (*Phragmidium mucronatum*) in his book 'Micrographia'. In 1835, Agostino Bassi, an Italian lawyer and farmer, first reported a fungal infection (muscardine) of silkworm and illustrated that a microbe can cause an infection. Robert Remak (1837–1841), a Polish physician, described the first human mycosis (tinea favosa). Gruby (1844) first described the aetiologic agent of tinea endothrix, later became known as *Trichophyton tonsurans*. The work of Remak and Gruby established the mycology as a separate branch of medical science. In 1892, Alejandro Posadas, a medical student, and Robert Wernicke, a pathologist, first described *Coccidioides* from a soldier with recurrent skin tumours in Argentina.

I. Samanta, *Veterinary Mycology*, DOI 10.1007/978-81-322-2280-4_1,

The histoplasmosis in human was first notified by Darling (1906), an American pathologist who observed it during an autopsy of a Martinique native person who died with tuberculosis-like syndrome in Panama. Raymond Jacques Adrien Sabouraud (France) compiled the description of dermatophytes (*Trichophyton*) in his book *Les Teignes* (1910) which was based on his observation in artificial culture. This authentic book initiated the development of medical mycology.

Gomori (1946) first developed a stain for the microscopic observation of fungal cells in tissue which was later modified by Grocott in 1955. Kligman (1951) used the periodic acid Schiff stain for histological demonstration of fungi. Gridley (1953) later modified PAS stain by replacing periodic acid with chromic acid which can reveal both hyphae and yeast cells in tissues.

In India, Lt. Col. Kirtikar (1885) established the mycology with his collection and identification of mushrooms from West Bengal. Cunningham (1872) exposed the grease covered slides to the air and collected spores of *Rhizopus* and rusts along with the cholera bacteria. Sir Edwin John Butler (1901) started the systemic study of fungi (*Pythium*, *Phytopthora*) and he is known as the 'Father of Indian mycology'.

1.2 Veterinary Mycology

In 1749, Reaumur first described avian aspergillosis in birds which was followed by the report of similar syndrome in a duck (Montagu, 1813). *Aspergillus fumigatus* was first detected in the lung of a great bustard (*Otis tardaga*) in 1863 by Fresenius. He was also the first to use the term 'aspergillosis' for this respiratory disease.

The causative agent of Epizootic lymphangitis in horses (*Histoplasma capsulatum* var. *farciminosum*) was first demonstrated in the pus by Rivolta (1873). However, it was successfully isolated in 1896 by Tokishiga in Japan. Smith (1884), a Veterinarian working in India, first described a chronic cutaneous granulomatous disease in horses known as 'bursattee' in local Indian language. Later the aetiology of the disease was correlated with the fungi (*Pythium*) based on histology.

2 General Characteristics of Fungi

2.1 Morphology

The fungi are eukaryotic, heterogeneous, unicellular to filamentous, spore bearing, and chemoorganotrophic organisms which lack chlorophyll. The fungi have three major morphological forms, i.e. unicellular yeast, filamentous mould (mold) and yeast-like form (pseudohyphae form). The dimorphic fungi (*Blastomyces dermatitidis*, *Coccidioides immitis*, *Histoplasma*, *Sporothrix schenckii*) are able to produce both the forms (yeast and mould) depending on the temperature (thermal dimorphism). The yeast form is produced within the body of the host (in vitro at 37 °C) and the mould form is observed either in the environment or in artificial culture medium (at room temperature). The pseudohyphae form is chains of elongated ellipsoidal cells with constriction between them and it is produced by *Candida albicans*.

Yeast: Yeasts are unicellular, microscopic but larger (1–5 μm × 5–30 μm) than most of the bacteria (0.5–1 μm). They are pleomorphic and show spherical, elliptical and elongated forms. In artificial culture media, the yeasts produce bacteria-like colonies which are moist or mucoid. The yeasts reproduce by budding. Both in the yeast and pseudohyphae form, the nuclear division and septa formation take place near the bud. The buds can elongate and are released as blastospores. Sometimes the buds elongate but fail to detach and they produce a chain of elongated hyphae-like filament called 'pseudohyphae'. The examples of pathogenic yeasts are *Cryptococcus neoformans* and *Candida albicans*. They are currently not considered as true yeast because they can exist in other forms such as pseudohyphae and hyphae (mycelial).

Mould: The whole body of a mould is called a thallus (plural, thalli). The thallus is structurally divided into two major parts, i.e. vegetative (mycelium) and reproductive portion (spores). The mycelium is composed of filaments with or without branching known as hypha (plural, hyphae). Each hypha (5–10 μm wide) is a tube-like structure containing a lumen surrounded externally by a rigid cell wall. The individual hyphae are intertwined to make a mycelium. Any part of the mycelium can absorb food. However, a specialised root-like structure (rhizoid) attaches the substrate to absorb the food in some group of fungi (Zygomycetes).

The content of the hyphal lumen is protoplasm. A bilayer membrane (plasma membrane or plasmalemma) separates the protoplasm from the outer cell wall. The elongation of the hypha takes place near the tip ('apical extension') by transverse cross wall formation from the existing cell wall. The cross wall grows inward to form a septum. A central pore is present in the septum which allows the movement of protoplasm along with the nuclei between the cells (protoplasmic streaming).

I. Samanta, *Veterinary Mycology*, DOI 10.1007/978-81-322-2280-4_2,

2.1.1 Types of Hyphae

The following hyphal types are observed:

(a) Vegetative hyphae: They penetrate the artificial medium to absorb the nutrients.
(b) Aerial hyphae: They grow above the surface of the artificial medium.
(c) Reproductive (fertile) hyphae: They are aerial hyphae carrying the reproductive structure (spore).
(d) Coenocytic hyphae: Non-septate hyphae which allow uninterrupted flow of protoplasm and nuclei through the lumen, e.g. Phycomycetes.
(e) Septate with uninucleated cells.
(f) Septate with multinucleated cells (Fig. 2.1).

Several fungi (dermatophytes) produce hyphae with specific appearance which helps in their identification. The examples are mentioned below:

(a) Spiral hyphae: Spirally coiled hyphae (*Trichophyton mentagrophytes*)
(b) Pectinate body: Short, unilateral projections from the hyphae resembling teeth of a comb (*Microsporum audouinii*)
(c) Favic chandeliers (antler hyphae): Irregular projections of the hyphae that collectively resemble a chandelier or the antler of a deer (*Trichophyton schoenleinii*, *T. violaceum*)
(d) Nodular body: Closely twisted hyphae resembling a nodule (*Trichophyton mentagrophytes*, *Microsporum canis*)
(e) Racquet hyphae: Chain of elongated hyphal cells expanded at one end to produce a tennis racquet-like arrangement (*Epidermophyton floccosum*, *Trichophyton mentagrophytes*) (Fig. 2.2)

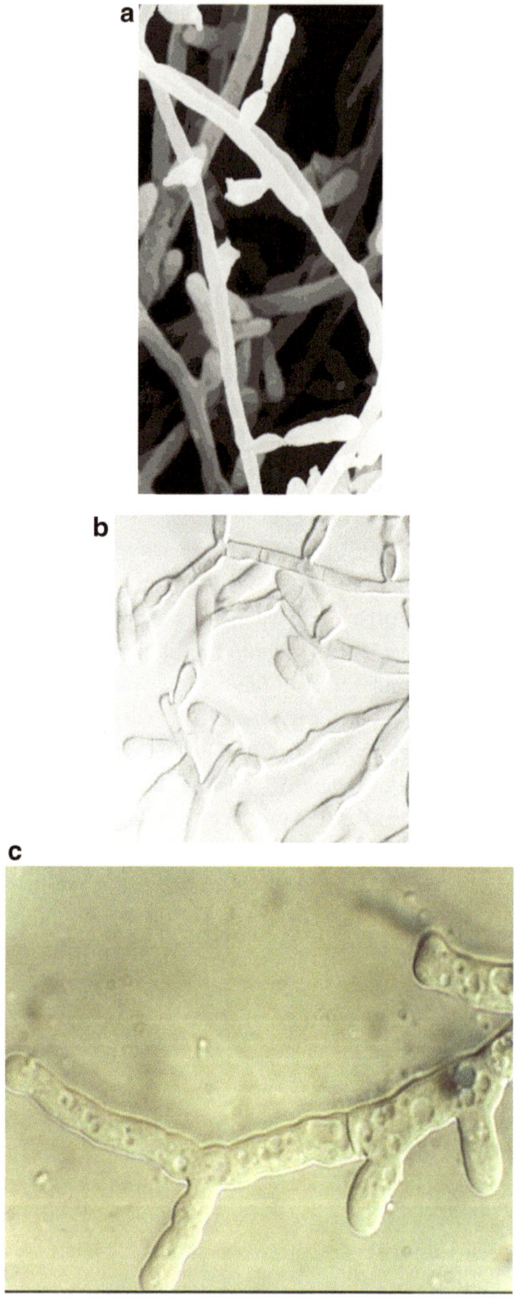

Fig. 2.1 Types of fungal hyphae; (**a**) coenocytic hyphae; (**b**) septate hyphae; (**c**) septate, multinucleate hyphae of *Penicillium* (Photograph courtesy: Prof. Sybren de Hoog and Dr. Marco van den Berg)

2.1.2 Fungal Cell Structure

Cell wall: All the morphological forms (yeast and hyphae) of fungi are surrounded by a rigid cell wall. The major component of the cell wall is chitinous fibrils embedded in a matrix of polysaccharides, proteins (acid phosphatase, α-amylase, and protease), lipids and inorganic salts (calcium, magnesium, phosphorus). Chitin is a (β1, 4) linked polymer of N acetyl-D-glucosamine (GlcNAc).It is synthesised by chitin synthetase present in the chitosome (cell organelle). The matrix contains glucan (D-glucose polymer),

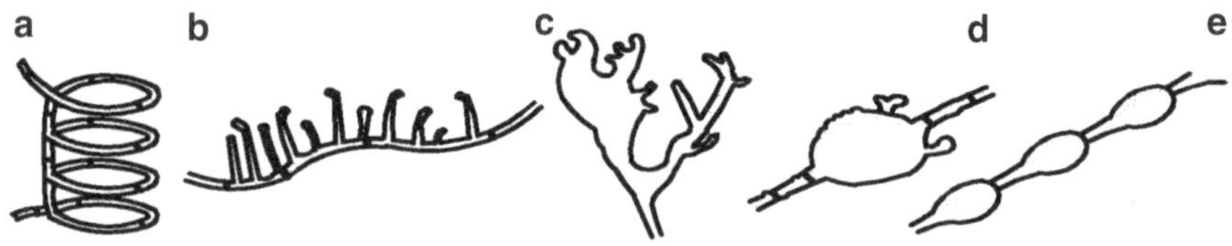

Fig. 2.2 Types of dermatophyte hyphae (schematic); *a*, spiral; *b*, pectinate body; *c*, favic chandelier; *d*, nodular organ; *e*, racquet hyphae

mannan, chitosan (polymer of glucosamine) and galactans. Among the glucans, polymers with (β1, 3) and (β1,6) linked glucosyl units, known as β glucan, is common. The β glucans are potent immunomodulator and are used in poultry industry.

In yeast form of dimorphic fungi (*Histoplasma*, *Blastomyces dermatitidis*), α glucan makes the cell layer, whereas the β glucan predominantly make the cell wall of mycelium. The presence of α glucan is associated with virulence of the yeast cells than the mycelial form. Another major constituent of yeast cell wall is peptidomannan (mannan, galactomannans, rhamnomannans) which helps in serological identification of the fungi. The capsule is observed in *Cryptococcus* like prokaryotes. The capsule consists primarily of two polysaccharides, glucuronoxylomannan (GXM) and glucuronoxylomannogalactan (GXMGal), along with smaller amounts of mannoproteins. It is antiphagocytic and it protects the yeast cells.

Plasma membrane: The plasma membrane is present beneath the cell wall. Like other mammalian cells, it is a phospholipid and sphingolipid bilayer in which the proteins (peripheral and integral) are interspersed. The hydrophilic heads of phospholipids are towards the surface, and the hydrophobic tails are present in the interior of the membrane. Major sterol of fungal plasma membrane is ergosterol. In contrast, mammalian plasma membrane contains cholesterol. The antifungals act on ergosterol (polyene) or biosynthetic pathway of ergosterol (imidazole, triazole) to inhibit the fungal growth.

Cytoplasm: The cell cytoplasm contains nucleus and organelles such as mitochondria, endoplasmic reticulum, golgi vesicles, lysosomes, vacuoles, etc. Long, hollow cylindrical structures (25 nm in diameter), known as microtubules, are also present in the cytoplasm. The microtubules are composed of tubulin protein and they help in the movement of chromosomes (during mitosis or meiosis), nuclei, golgi vesicles, vacuoles and mitochondria. Disruption of tubulin synthesis by antifungal (griseofulvin) can prevent the fungal mitosis. The cytoplasm contains membrane-bound nucleus composed of chromatin and nucleolus. The chromatin is composed of DNA and associated proteins. The number, shape and size of nuclei vary between the fungi. Majority of genetic material (80 %) is present in the chromatin and the rest (20 %) is associated with mitochondrial genome (Fig. 2.3).

2.2 Nutrition and Growth

The fungi are chemoorganoheterotrophic organisms. They use chemical compounds as a source of energy and organic compounds as electron and carbon source. They obtain their nutrition by absorption (osmotrophic) either from the environment (saprophyte) or the host (parasite). Most of the saprophytic moulds grow aerobically in artificial culture medium at 20–30 °C. The pathogenic yeasts and yeast phase of dimorphic fungi prefer to grow at 37 °C. High humidity, acidic pH (3.8–5.6), high sugar concentration (4–5 %), carbon, phosphorus, sulphur and traces of potassium, magnesium, iron and calcium are required for optimum fungal growth. The peptone in the media and keratin in the skin act as a nitrogen source. The nitrogen is required for synthesis of amino acids for building proteins, purines and pyrimidines for nucleic acids, glucosamine for chitin and various vitamins. Most of the fungi use nitrogen as nitrate which is reduced to nitrite and further to ammonia. None of them can directly fix

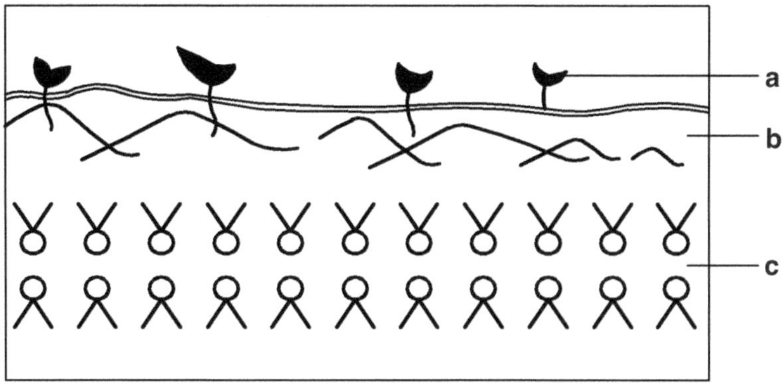

Fig. 2.3 Fungal cell wall (schematic); *a*, protein; *b*, glucan; *c*, cytoplasmic membrane

Table 2.1 Comparative growth requirements of bacteria and fungi

Characteristics	Bacteria	Fungi
Temperature	37 °C	20–30 °C (saprophyte)/37 °C (parasite)
pH	6–7	3.8–5.6
Oxygen	Aerobic, anaerobic	Strictly aerobic
Sugar	0.5–1 %	4–5 %
Humidity	Low	High
Carbon	Organic or inorganic	Organic
Nitrogen	Direct fixing by some bacteria (*Rhizobacter*)	No direct fixing
Lysine synthesis	*meso*-α,ε-diaminopimelic acid (DAP pathway)	L-α-adipic acid (AAA pathway)
Growth rate	Fast	Slow

nitrogen. The growth is not dependent on light and ultraviolet and X-ray are mild inhibitory. The growth rate of fungi is slower than bacteria and the medium is easily contaminated with bacteria. Antibiotics (e.g. chloramphenicol) and antifungal (e.g. cycloheximide) are added in the media to prevent the bacterial and saprophytic fungi contamination. The cycloheximide is inhibitory against the growth of certain pathogenic fungi and yeasts such as *Aspergillus fumigatus*, *Candida albicans* and *Cryptococcus*.

Inoculation of culture medium is done by a transfer needle with flattened tip which helps in cutting of mycelium. The bacteriological inoculation loop is sufficient for inoculation of yeasts (Table 2.1).

2.3 Reproduction

The fungi reproduce by two major ways, i.e. asexual and sexual reproduction. The reproduction of fungi produces spores which are considered as dispersal and survival unit of fungi without an embryo. The spores separate from their mother fungi and develop into an individual progeny.

In asexual reproduction, union of sex organ or nuclei does not occur. Instead, the following processes are observed:

1. Fission of mother cell: It produces two similar daughter cells.
2. Budding: The budding has three steps:
   (i) Bud emergence: The outer cell wall of the parent cell thins and new inner cell wall and plasma membrane are synthesised at the site of bud emergence. The bud emergence is regulated by the activation of the polysaccharide synthetase (zymogen) for the synthesis of new cell wall and turgor pressure of the parent cell.
   (ii) Bud growth: Mitosis occurs and the conidium (spore) possesses the daughter nucleus.

(iii) Conidium separation: The septum between the developing conidium and its parent cell is formed from a chitin ring. The septum helps in the separation of conidia which leaves a bud scar on the parent cell wall. Sometimes the conidia are not separated and released from the parent cells. Consequently, pseudohypha consisting of a filament of attached conidia is produced (Fig. 2.4).

3. Fragmentation of hyphae.
4. Spore formation.

Many kinds of fungal asexual spores are detected.

(a) Sporangiospores: A specialised reproductive hypha (sporangiophore) bears a sac-like structure known as sporangium. Each of the sporangium contains numerous single-celled sporangiospores. Non-motile sporangiospores are called 'aplanospore' and motile sporangiospores are called 'zoospore'.
(b) Conidia: Conidia are small asexual spores attached directly with reproductive hyphae (conidiophore). Small single-celled spores are called microconidia and large multicellular spores are called macroconidia.

Fig. 2.4 Steps of yeast budding (schematic); *a*, bud emergence; *b*, bud growth; *c*, conidium separation

(c) Arthrospores (oidia): Disjoining of hyphal cells produces single-celled arthrospores.
(d) Chlamydospores: Thick-walled, single-celled spores produced by aerial hyphae and they are highly resistant to adverse environment. Some fungi (*Histoplasma*) produce chlamydospores with small spine-like projections in their wall (tuberculate).
(e) Blastoconidia: The spores produced by budding are known as blastoconidia (Fig. 2.5).

Sexual reproduction occurs by the union of two compatible thalli (homothallic or heterothallic).In heterothallic fungi the male and female mating types exist in nature, designated as **A** and **a** (or + and −). The homothallic fungi do not require separate mating types. The sex organelle of fungi is called gametangium. Male and female gametangia are known as 'antheridium' and 'oogonium', respectively. The union of compatible thalli is followed by fusion of gametangia (plasmogamy). In some fungi, 'trichogyne' (a specialised hyphae) is present surrounding the female gametangium and it receives the male gamete. The male gamete is uninucleate spermatium (microconidium) or multinucleate macroconidium, formed either directly on the male mycelium (+ type) or in a specialised structure called 'spermogonia'. The trichogyne recruits a fertilising nucleus from the male gamete which enters the female gametangium and fusion of two haploid nuclei (karyogamy) occurs. The karyogamy produces a diploid nucleus which undergoes meiosis to reduce the chromosome number into haploid again. Some of the progeny nuclei degenerate and in some fungal species post-meiotic mitosis occurs. Sexual reproduction is typically controlled by genes

Fig. 2.5 Types of fungal asexual spores (schematic); *a*, sporangiospore; *b*, conidia; *c*, arthrospores; *d*, chlamydospores; *e*, blastospores

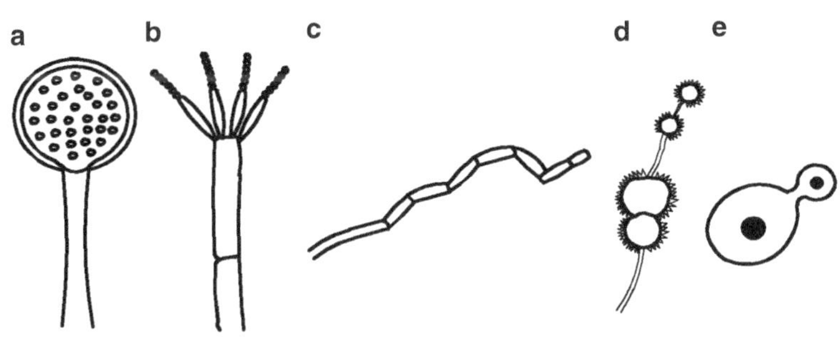

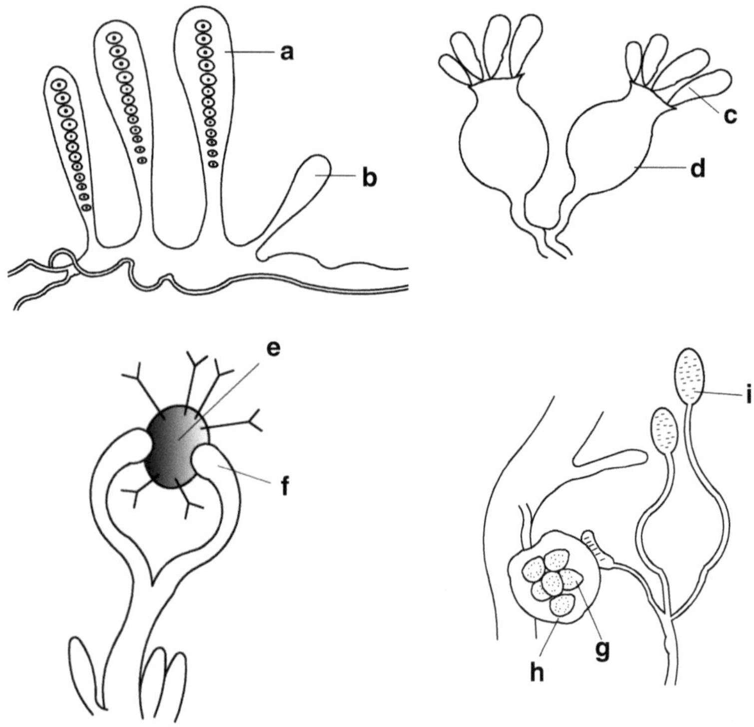

Fig. 2.6 Types of fungal sexual spores (schematic); *a*, ascospore; *b*, ascus; *c*, basidiospore; *d*, basidium; *e*, zygospore; *f*, gametangium; *g*, oospore; *h*, oogonium; *i*, antheridium

that reside in the mating-type locus. The sexual spores occur less frequently and in smaller numbers than the asexual spores. Different types of sexual spores are described below.

(a) Ascospores: These meiospores are single celled and are produced within a sac-like structure, known as ascus (meiosporangium). Each ascus contains eight ascospores.
(b) Basidiospores: These spores are also like ascospores but are produced within a club-shaped structure known as basidium.
(c) Zygospores: Zygospores are produced by the fusion of two compatible thalli or their gametangia.
(d) Oospores: These special sexual spores are produced within female gametangium (oogonium). After fusion of two compatible hyphae, the male gametes (produced in antheridium) enter the oogonium and fertilise the female gametes (oospheres) to produce oospores (Fig. 2.6).

The sexual and asexual spores are protected from the environment by a highly organised structure, known as 'fruiting body' or 'sporocarp'. The fruiting bodies for sexual spores include cleistothecia, perithecia, apothecia and pseudothecia. The cleistothecium is a completely closed fruiting body. In perithecium, a pore (ostiole) is present through which the spores can escape. The apothecium is a cup or saucer-like fruiting body. In addition to spores, some fruiting bodies contain sterile hyphae.

The fruiting bodies such as pycnidium and acervulus can protect asexual spores. Pycnidium is a spherical structure with a pore (ostiole) at the top and it is arranged as separate or aggregate. The acervulus is an open, saucer-shaped asexual fruiting body.

3 Classification of Fungi

The classification of fungi relies mostly on morphological criteria such as the pigmentation, shape of hyphae, presence or absence of septa and types of spores. The taxonomy of mould and yeasts is governed by International Code of Botanical Nomenclature (ICBN). Any new proposal for classification of fungi is published in official journal of International Association for Plant Taxonomy (*Taxon*) and is discussed in annual meeting before acceptance. The classification of clinically relevant fungi is described in Table 3.1.

I. Samanta, *Veterinary Mycology*, DOI 10.1007/978-81-322-2280-4_3,

Table 3.1 Classification of clinically relevant fungi

Kingdom	Phylum	Class	Order	Family	Genus	Species
Fungi	Zygomycota	Zygomycetes (previously Phycomycetes)	Entomophthorales	Basidiobolaceae	*Basidiobolus*	*Basidiobolus ranarum*
				Ancylistaceae	*Conidiobolus*	*Conidiobolus coronatus*
	Basidiomycota	Tremellomycetes	Tremellales	Tremellaceae	*Cryptococcus*	*Cryptococcus neoformans* var. *neoformans*, *C. neoformans* var. *grubii*, *C. gattii*
	Ascomycota	Euascomycetes	Onygenales	Ajellomycetaceae	*Blastomyces*	*Blastomyces dermatitidis*
				Onygenaceae	*Histoplasma*	*Histoplasma capsulatum*
					Coccidioides	*Coccidioides immitis*, *C. posadasii*
					Emmonsia	*E. crescens*, *E. parva*
				Arthrodermataceae	*Trichophyton*	*T. tonsurans*, *T. equinum*, *T. interdigitale*
					Microsporum	*M. audouinii*, *M. canis*, *M. gypseum-fulvum* complex
					Epidermophyton	*E. floccosum*, *E. stockdaleae*
					Chrysosporium	
		Eurotiomycetes	Microascales	Microascaceae	*Aspergillus*	*A. fumigatus*, *A. flavus*, *A. terreus*, *A. niger*, *A. nidulans*, *A. ustus*
			Eurotiales	Trichocomaceae	*Penicillium*	*P. marneffei*
		Pyrenomycetes	Ophiostomatales	Ophiostomataceae	*Sporothrix*	*S. schenckii*
		Saccharomycetes (Hemiascomycetes)	Saccharomycetales	Saccharomycetaceae	*Candida*	*C. albicans*, *C. tropicalis*, *C. glabrata*
		Dothideomycetes	Capnodiales	Davidiellaceae	*Cladosporium*	*C. cladosporioides*
			Pleosporales	Pleosporaceae	*Alternaria*	*A. alternate*
					Bipolaris	*B. spicifera*
		Eurotiomycetes	Chaetothyriales	Herpotrichiellaceae	*Cladophialophora*, *Exophiala*, *Fonsecaea*, *Phialophora*	*C. bantiana*, *E. attenuate*, *F. multimorphosa*, *P. verrucosa*
			Ochroconiales		*Ochroconis*	*O. gallopava*
		Archiascomycetes	Pneumocystidales	Pneumocystidaceae	*Pneumocystis*	*P. carinii*
	Mucoromycotina (subphylum) (*incertae sedis*, not assigned to any phylum)		Mucorales	Mucoraceae	*Rhizopus*, *Absidia*, *Mucor*	*Rhizopus arrhizus* (*R. oryzae*), *Mucor circinelloides* complex
Stramenopila (Chromista)	Oomycota	Oomycetes	Pythiales	*Pythiaceae*	*Pythium*	*P. insidiosum*
					Lagenidium	*L. giganteum*

4 Cutaneous, Subcutaneous and Systemic Mycology

4.1 *Trichophyton*

The first description of dermatophytosis was recorded by Celsus, a Roman encyclopaedist who described a suppurative infection of scalp ('porrigo' or 'kerion of Celsus') in *De Re Medicina* (30 A.D.). Throughout the middle ages, several descriptions of dermatophytosis were produced where it is described as 'tinea'. The keratin-destroying moths which made circular holes in the woollen garments are known as *Tinea*. Due to similarity in the structure of circular lesion of dermatophytosis on the smooth skin with the circular hole made by moth, Cassius Felix introduced the term 'tinea' to describe the lesions. In 1806, Alibert used the term 'favus' to describe the honey-like exudate in some scalp infections. However, the fungal aetiology of tinea was first detected by Robert Remak, a Polish physician who first observed the presence of hyphae in the crusts of favus. This detection is also a landmark in medical history because this is the first description of a microbe causing a human disease. He himself did not publish his work, but he permitted the reference of his observations in a dissertation by Xavier Hube in 1837. Remak gave all the credits of his discovery to his mentor Schoenlein who first published the fungal etiological report of favus in 1839. He observed the infectious nature of the favus by autoinoculation into his own hands and also successfully isolated the fungus later (1945) and named *Achorion schoenleinii* (*Trichophyton schoenleinii*) in honour of his mentor. In 1844, Gruby described the etiologic agent of tinea endothrix, later became known as *Trichophyton tonsurans*. The genus *Trichophyton* was created and described by Malmsten (1845) with its representative species *T. tonsurans*. Charles Robin identified *T. mentagrophytes* in 1847 and *T. equinum* was identified by Matruchot and Dassonville in 1898. Raymond Jacques Adrien Sabouraud (France) first compiled the description of *Trichophyton* in his book (*Les Teignes*) in 1910 which was based on his observation in artificial culture. The sexual state of dermatophyte was described by Nannizzi (1927). Emmons (1934) first reported the classification of dermatophytes based on vegetative structures and conidia. Gentles (1958) established the successful treatment of tinea capitis with griseofulvin.

In India, Tewari (1962) first isolated *Trichophyton verrucosum*, *Trichophyton mentagrophytes* and *Trichophyton violaceum* from calves, dogs and poultry. *T. rubrum* was first isolated from dog and calves at Bengal Veterinary college, Kolkata (currently West Bengal University of Animal and Fishery Sciences) (Chakrabarty et al. 1954). Tewari (1969) isolated *T. simii* from chicken, dog and man for the first time in India. *Trichophyton verrucosum* infection in camel and its handlers is also reported from India (Pal and Lee 2000).

I. Samanta, *Veterinary Mycology*, DOI 10.1007/978-81-322-2280-4_4,

4.1.1 Morphology

The septate, branching hyphae are produced by *Trichophyton* in the artificial culture and non-parasitic (environmental) state. Sometimes, abnormal forms of hyphae such as spiral or coiled hyphae, tennis racquet hyphae, pectinate hyphae (like teeth of a comb) and 'favic chandeliers' (irregular projections along one side of the hyphae) are observed (Fig. 4.1). The hyphae contain two genetically distinct nuclei (heterokaryon). So, the ploidy of the nucleus is uncertain.

The hyphae produce asexual spore known as conidia which generally contains single nucleus. Two types of conidia i.e. macroconidia (macroaleuriospores) and microconidia (micro-aleuriospores) are detected. The macroconidia are large (up to 100 μm long), smooth-walled, slender club-shaped, blunt at the end, and with numerous transverse septa. Their presence is variable, whereas the microconidia are abundant in presence, small, thin walled, hyaline, sub-spherical to club-shaped and borne singly or in grape-like cluster (Fig. 4.2). In the parasitic state, *Trichophyton* produces hyphae and arthroconidia.

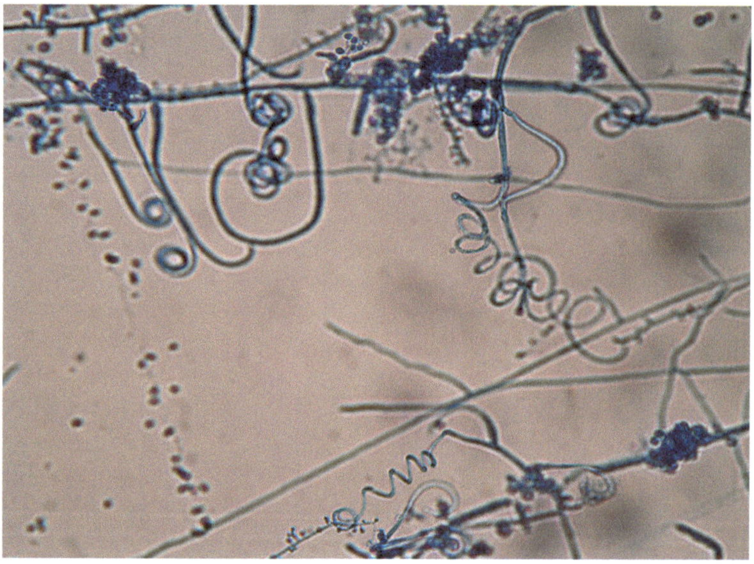

Fig. 4.1 Coiled hyphae of *Trichophyton mentagrophytes* (Photograph courtesy – Prof. Alexandro Bonifaz, Head, Department of Mycology and Dermatology service, General Hospital of Mexico, Mexico City)

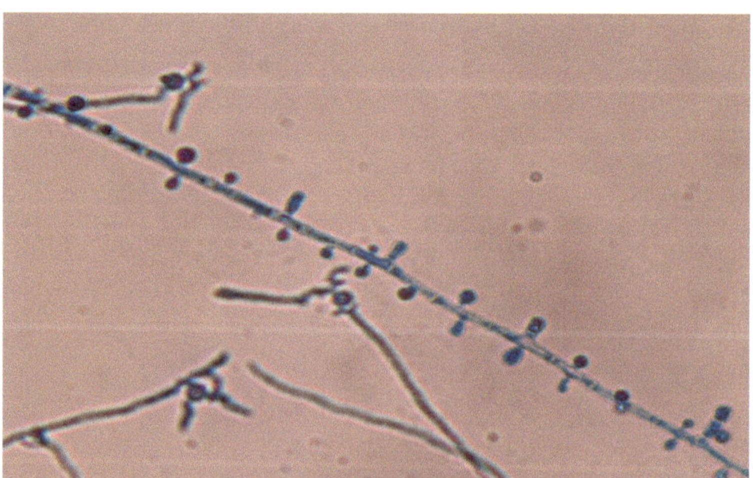

Fig. 4.2 Microconidia of *Trichophyton rubrum* (Photograph courtesy – Prof. Alexandro Bonifaz, Head, Department of Mycology and Dermatology service, General Hospital of Mexico, Mexico City)

4.1.2 Classification

All the dermatophytes are classified into three ecological groups namely geophilic (soil), zoophilic (animals) and anthropophilic (human). The geophilic dermatophytes are saprophytes and they derive their nutrients from keratinous substrates (hair, feathers, horns) present in the environment. They may infect human and animals during direct contact with the contaminated soil. The examples include *Trichophyton ajelloi* and *Trichophyton terrestre*. The zoophiles have a specific animal host and they can infect human also. The examples include *Trichophyton simii* (monkeys), *Trichophyton mentagrophytes* (rodents) and *Trichophyton equinum* (horses), whereas the host of anthropophilic species are human, but they can infect animals also. The examples of anthropophilic species are *Trichophyton rubrum*, *T. kanei*, *T. schoenleini*, *T. concentricum* and *T. tonsurans*.

All dermatophytes belonged to the phylum Ascomycota, class Euascomycetes, order Onygenales and family Arthrodermataceae. The family Arthrodermataceae contains four morphological anamorph genera such as *Trichophyton*, *Microsporum*, *Epidermophyton* and *Chrysosporium*. The molecular study revealed that none of the genera are monophyletic. The anthropophilic and zoophilic species in the genus *Trichophyton* are present in a single clad, whereas the geophilic species are present in another clad. It seems that the ecological niche acted as a motivational force for evolution of *Trichophyton*. Further, within the same species complex such as *Trichophyton mentagrophytes* complex, both anthropophilic (*T. mentagrophytes* var. *interdigitale*) and zoophilic (*T. mentagrophytes* var. *mentagrophytes*) species are present.

The *Trichophyton* genus contains 25 species arranged under different teleomorphic species complex. The type species is *Trichophyton tonsurans*. The teleomorphic species complexes are *Arthroderma vanbreuseghemii* complex (*T. tonsurans*, *T. equinum*, *T. interdigitale*), *Arthroderma simii* complex (*T. simii*, originally described from India and associated with monkeys), *Arthroderma benhamiae* complex (*T. mentagrophytes* var. *granulosum*, *T. mentagrophytes* var. *erinacei*) and *Trichophyton rubrum* complex (*T. rubrum*, *T. violaceum*). The details of the anamorph species and their hosts under these complexes are described in Table 4.1.

Table 4.1 Teleomorphic species complex, anamorph species and hosts of *Trichophyton*

Teleomorphic species complex	Anamorph	Host
Trichophyton rubrum complex	*T. rubrum* of Africa (including the old taxa of *T. raubitschekii*, *T. soudanense*, *T. gourvilii*, *T. megninii*)	Human
	T. rubrum (including the old taxa of *T. fischeri*, *T. kanei*, *T. raubitschekii*)	Human
	T. violaceum (including the old taxa of *T. yaoundei*, *T. glabrum*)	Human
Arthroderma simii complex	*T. simii*	Monkey
	T. mentagrophytes (including the old taxa *T. mentagrophytes* var. *quinckeanum*, *T. langeronii*, *T. sarkisovii*)	Mice, camels
	T. schoenleinii	Human
Arthroderma benhamiae complex	*T. verrucosum*	Cattle
	T. erinacei	Hedgehogs
	Trichophyton species of *A. benhamiae* (including *T. mentagrophytes* var. *granulosum*)	Rabbits, guinea pigs, rodents
	T. concentricum	Human
	T. bullosum	Horses
	T. eriotrephon	–
Arthroderma vanbreuseghemii complex	*T. tonsurans*	Human
	T. equinum	Horses
	T. interdigitale (including zoophilic species as *T. mentagrophytes* var. *mentagrophytes*, var. *granulosum*, T. *verrucosum* var. *autotrophicum*)	–
	T. mentagrophytes var. *goetzii*, *interdigitale*, *nodulare*, *T. krajdenii*	Human

Table 4.2 Teleomorphs of *Trichophyton*

Anamorph	Teleomorph
T. mentagrophytes var. *mentagrophytes*	*A. benhamiae*
T. georgiae	*A. ciferrii*
T. flavescens	*A. flavescens*
T. vanbreuseghemii	*A. gertleri*
T. gloriae	*A. gloriae*
T. terrestre	*A. insingulare*
T. terrestre	*A. lenticularum*
T. terrestre	*A. quadrifidum*
T. simii	*A. simii*
T. mentagrophytes var. *interdigitale*	*A. vanbreuseghemii*
Unnamed	*A. curreyi*
Unnamed	*A. tuberculatum*

4.1.3 Reproduction

The sexual reproduction cycle is observed in geophilic and zoophilic (except *T. equinum*) species of dermatophytes, whereas the anthropophilic species do not possess such kind of sexual cycle and probably they reproduce clonally. The mating competent species have an asexual type (anamorph) and a sexual type (teleomorph). The sexual stage (teleomorph) is known as *Arthroderma* (Table 4.2).

The sexual reproduction of other ascomycetous fungi such as *Aspergillus* is controlled by *MAT* locus with two idiomorphs such as *MAT-1* and *MAT-2*. The heterothallic *MAT-1* and *MAT-2* can undergo sexual mating. The similar kind of *MAT* locus is recently identified in three species of *Trichophyton* i.e. *T. equinum*, *T. rubrum*, and *T. tonsurans* and two species of *Microsporum*. All of them shared certain common features in *MAT* locus such as small size (around 3 Kb) and specific arrangement of *SLA2*, *COX13*, and *APN2* genes at one side of the locus. In other Ascomycota these genes flank the *MAT* locus. The transcriptional orientations of *APN2* and *COX13* genes in dermatophytes are also different from other dimorphic fungi. Further, successful mating is observed between *T. rubrum* and *Arthroderma simii* which suggests the possibility of cross-species sexual recombination.

4.1.4 Susceptibility to Disinfectants

The experimental study revealed that among disinfectants such as chlorine, phenol, sodium dodecyl sulphate, and quaternary ammonium salts, chlorine (1 %) exhibited the strongest fungicidal action against *Trichophyton* inoculum.

4.1.5 Natural Habitat and Distribution

The geophilic species of *Trichophyton* prefers to reside in the soil. They can produce the antibacterial substances to survive from the soil bacteria. The zoophilic and anthropophilic species can use the host keratin and prefer to reside the hair, fur, and feathers of human or animals. They can reside in the soil if embedded in skin scales, hair and feathers. It occurs in animal sheds where recurrent infection is common. The conidia can survive in the environment with high moisture and low temperature. The conidia also remain alive in the skin and hair after the recovery of the animals.

Trichophyton is worldwide in distribution. The distribution matrix of different species changes with time. Before the Second World War, prevalence of *T. mentagrophytes* and *T. verrucosum* was highest in human causing onychomycosis and tinea pedis in North and Central Europe. The prevalence of *T. rubrum* increased steadily among the human population after 1948–1950. In Southern Europe and Arabic countries, zoophilic dermatophytes, such as *T. verrucosum* are the most frequently isolated species. In the United States and Mexico, frequency of *T. rubrum* isolation from human is followed by *T. mentagrophytes* and *T. tonsurans*, whereas *T. simii* was proposed to be endemic in Indian subcontinent as it was commonly isolated from soil, poultry and other animals in India. Recently it is also isolated from non-endemic areas such as Belgium, France and Ivory Coast. In India, other studies indicated the prevalence of *T. mentagrophytes* var. *mentagrophytes* and *T. rubrum* in pet dogs which was further transmitted to their owners.

4.1.6 Genome

Currently whole genome sequences of four *Trichophyton* species (*T. equinum*, *T. rubrum*, *T. tonsurans*, and *T. verrucosum*) are available at Broad Institute's dermatophyte comparative database (http://www.broadinstitute.org). The dermatophyte genome is enriched in the InterPro domains (protease, kinase, secondary metabolite, LysM) and it contains 308 genes specific to the dermatophytes. Further, species-specific genes are also present which play major role in host preference.

The size of *T. rubrum* genome is 22 Mb and it contains five chromosomes of 5.8, 5.2, 4.6, 3.05, and 3.0 Mb sizes. The repetitive DNA constitutes 5–10 % of the genome, whereas the size of *T. verrucosum* genome is 22.6 Mb containing 8,024 predicted protein-encoding genes. Approximately 5744 genes of *T. verrucosum* genome contain introns. The GC content is below 40 %. The GC rich region is present as 'long mosaic' in the genome separated by 'AT islands'. The genome encodes the enzymes for glycolysis, tricarboxylic acid cycle, pentose phosphate shunt, and synthesis of all amino acids and the nucleic acid bases, proteases (235 protease encoding genes are present). The number of tRNA genes is low (77) in comparison to other fungi.

4.1.7 Isolation, Growth and Colony Characteristics

The standard isolation medium is dermatophyte test medium (DTM) containing antifungals and antibiotics such as cycloheximide, chloramphenicol and gentamicin that can inhibit the growth of contaminant fungi and bacteria. Some species have fastidious requirements. *T. equinum* requires nicotinic acid and *T. verrucosum* requires thiamine and inositol for their optimum growth. The DTM also contains phenol red indicator which changes the colour of the medium from yellow to red during growth of *Trichophyton*. The proteolysis occurs during the growth which liberates ammonium ion and the pH of the medium changes to alkaline.

The conventional solid medium includes Sabouraud dextrose agar with cycloheximide, chloramphenicol and gentamicin. For detection of sporulation, potato dextrose agar is preferred. *T. schoenleinii* are isolated in Sabouraud glucose agar, BCP milk solid glucose agar. The cultures are incubated at 25–30 °C except *T. verrucosum* which requires 37 °C temperatures for optimum growth. The incubation period is 2 weeks and the absence of growth after 3–6 weeks is considered as negative.

On Sabouraud dextrose agar, *T. verrucosum* colonies are slow growing, small, disc-shaped, white to cream coloured, with a velvety surface, a raised centre and flat periphery. The colony surface pigmentation is white-grey, pink, buff or pale yellow and the reverse pigment may vary from non-pigmented to pink/yellow. The surface and reverse colour of the colonies varies with the species of *Trichophyton* which aids in diagnosis.

Extensive pleomorphism is observed in *T. mentagrophytes* cultures. The pleomorphism converts the morphology of fungal colony from granular texture to white fluffy tufts of aerial mycelium on the surface of colonies which results in the loss of characteristic pigmentation and conidia production.

4.1.8 Antigenic Characteristics

The immunodominant antigens of *Trichophyton* are expressed during sporulation and early hyphal growth. Two types of keratinases are secreted by them which belonged to subtilase (serine proteases) and zinc metalloprotease family. The recombinant subtilase enzyme of *T. rubrum* (Tri r 2) is able to stimulate immediate and delayed-type hypersensitivity. The P5 peptide is identified as immunodominant epitope associated with delayed-type hypersensitivity. The heat shock protein (hsp) of *T. mentagrophytes* is used to formulate plasmid DNA-based vaccine in laboratory animal model which offers some degree of protection. It indicates the immunodominance nature of *T. mentagrophytes* hsp.

Table 4.3 Virulence mechanisms and factors possessed by *Trichophyton*

Virulence factors	Functions
Endoprotease	They help to degrade the keratin
(a) Subtilisins (sub, alkaline proteases, pH 7–9 is required for optimum activity)	
(b) Fungalysins (Mep, Zn metalloproteases with HEXXE motif. Their optimum pH of activity is between 7.0 and 8.0.)	
Exoproteases	The Dpp and Lap synergistically can degrade the large peptides into oligopeptides and free amino acids
(a) Serine proteases (DppIV, DppV)	
(b) Leucine aminopeptidases (Lap1, Lap2)	
(c) Carboxypeptidases (McpA, ScpA)	
Kinase	The kinases are involved in signalling necessary for adaptation of *Trichophyton* in the skin
Mannan	It can inhibit the macrophage-mediated phagocytosis of conidia and inhibit the cell-mediated immune response
Sulphite	In the presence of sulphite, disulphide bonds of the keratin substrate are directly cleaved to cysteine and S-sulphocysteine. The reduced proteins become accessible to other proteases for further cleavage
LysM	The LysM protein conceals the dermatophyte cell wall components and carbohydrates and thus avoids the host immune response to the fungi

4.1.9 Virulence Factors

The virulence factors of *Trichophyton* explored by different investigations till date are enlisted in Table 4.3.

4.1.10 Transmission

Most of the dermatophytes including *Trichophyton* are transmitted into the host through the injured skin, scars and burns. The arthrospores or conidia are the major infectious agents. The resting hairs do not possess essential nutrients for the dermatophytes, so the resting hairs are not infected with the organism. The infection after entry into a herd, kennel or cattery, spreads from one animal to another by direct transmission or indirectly via environmental contamination or fomites. In farms, high animal density and close contact promote direct transmission between animals in the herd or between the herds in pasture. Shedding of dermatophyte spores contaminates the farm environment which can last for several years depending upon the moisture. The rodents also help to spread the infection by leaving infected hairs around their barns.

The zoophilic *Trichophyton* (*T. verrucosum*, *T. equinum*) are transmitted from animals to human by direct contact with infected animals or through the environment contaminated with the animal fomites. This kind of skin infection is an occupational disease of farmers and their family members, Veterinarians, abattoir and tannery workers and is common in pet owners especially the children who handle the infected dogs and cats.

The human-to-human transmission occurs primarily through the indirect way such as loose strands of hair and desquamated epithelial cells (*T. schoenleinii*) than the direct contact. The transmission may occur via contaminated hairbrushes, combs and hats. Transmission of infection from mother to newborn child is possible. *T. schoenleinii* can persist in the houses for several generations without proper disinfection.

4.1.11 Pathogenesis

The dermatophytes do not survive in the live cells and areas of inflammation. So, they do not invade the skin epithelial cells. The spores, after transmission through the skin trauma, can germinate in the non-living keratinised layer of the skin (stratum corneum). The fungal metabolites induce inflammatory reaction at the site of infection. The inflammation causes the pathogen to move away from the site of infection in a circular way with healing at the centre and papules at the periphery which is responsible for classical ring shaped lesion.

The invasion of stratum corneum is associated with the production of different virulence factors such as sulphite, keratinase and proteases as described in Table 4.3. The keratinous tissues are composed of insoluble proteinaceous complex made of a network of different cross-linked proteins such as loricrin- and proline-rich proteins which contain numerous cysteine residues to form disulphide bridges. The secreted sulphite can break these disulphide bridges and make the reduced proteins accessible by different proteases and keratinases for further digestion.

The pathogenesis of anthropophilic *T. rubrum* causing Majocchi's granuloma in human is little different because the fungi can invade and grow in the moist living tissues. The trauma causing the breakage in the epidermis is a prerequisite for the establishment of the infection in the dermis. During trauma, follicular disruption occurs which causes the introduction of keratin and necrosed materials into the dermis. The keratin and necrosed materials act as substrate for survival of the organism. The growth of the organism causes cellular destruction and inflammation which raises the level of stromal acid mucopolysaccharides, lowering the dermal pH to make it suitable for the fungal survival. The neutrophils and monocytes can ingest and kill the conidia of *T. rubrum*. However during immunosuppression these neutrophils and monocyte mediated killing is hampered and the infection may progress into the deeper part of the tissues.

4.1.12 Disease Produced

The major animal and human diseases produced by *Trichophyton* are enlisted in Table 4.4.

Table 4.4 Major diseases of animal and human caused by *Trichophyton*

Fungi	Host	Disease
Trichophyton verrucosum, *T. mentagrophytes*, *T. rubrum*	Cattle	*Bovine ringworm*: The calves are most susceptible especially when kept within the crowded pens. The lesions are circular, painless, thick, white and scattered with occasional production of large plaques ('asbestos') with thick scabs and crusts. The lesions appear in the head, neck and less frequently in the back, flank and limbs. The secondary bacterial infection may occur associated with pruritis. Other clinical signs are skin scaling and loss of hair. Both *T. verrucosum* and *T. mentagrophytes* produce ectothrix infection with the formation of spores on the surface of the hairs
T. equinum	Horse	*Equine dermatophytosis*: The dry, scaly and multiple lesions appear in any part of the body especially in the groomed part. Sometimes the lesions become larger to produce a 'moth-eaten appearance'
T. verrucosum ('club-lamb fungus')	Sheep	*Ovine dermatophytosis*: The lesions are scaly and appear in the hairless part of the face, ear and neck or it may soil the wool. The infection is self-limiting, but the wool will not appear in the infected area for the next 4–5 months
T. simii	Poultry	The lesions are scaly, inflammatory and necrotizing which appear in featherless part of the body. The disease is endemic in India
T. mentagrophytes	Dogs and cats	The scaling to inflammatory lesions appears in the body with occasional secondary bacterial infection and suppuration. The later is known as 'wet eczema'. Affected area become hairless and the vesicles appear at the periphery

(continued)

Table 4.4 (continued)

Fungi	Host	Disease
T. rubrum, *T. mentagrophytes*	Human	*Tinea corporis* (*tinea circinata*): It is the classical ringworm infection in human which is characterised by the formation of lesion in trunk and extremities (non-hairy glabrous region of body). The lesions are pink-to-red annular or arciform patches with scaly or vesicular borders that expand peripherally with a tendency for central clearing. Sometimes follicular papules are present at the active border. When it happens in the face, it is known as *tinea faciei*, and when it affects the head, neck, upper extremities but rarely the legs, it is known as *tinea corporis gladiatorum*
T. concentricum	Human	*Tinea imbricata* (*Tokelau*): It is a superficial mycosis restricted to South Pacific islands (Polynesia and Melanesia), South Asia and South America. The skin lesions are multiple concentric annular plaques with thick adherent peripheral rims of scale
T. rubrum, *T. mentagrophytes*, *T. mentagrophytes* var. *interdigitale* (ulcerative and vesiculobullous)	Human	*Tinea pedis* (*athletes' foot*): It is the ringworm infection in which the lesions are found in feet. Clinically it has four types such as interdigital, moccasin, ulcerative and vesiculobullous. The erythema, scaling and maceration with fissures are observed in interdigital type which occurs in the space between 4th and 5th toes. The diffuse scaling on the soles extending to the sides of the feet is found in the moccasin type. In ulcerative type, lesions are macerated with scaling border. Secondary bacterial infection is a common feature in this type of tinea pedis. In vesiculobullous type, vesicular eruptions occur at the side of the feet
T. rubrum, *T. mentagrophytes*	Human	*Tinea manuum*: It is a diffuse hyperkeratosis which is commonly found in palmar creases of the palms and digits. White powdery scales along the palmar creases are typically seen. It may occur in one hand along with tinea pedis (two feet and one hand syndrome)
T. rubrum, *T. mentagrophytes*	Human	*Tinea cruris (Jock's itch)*: The lesions are found in the crural folds which may extend to the thighs, buttocks and gluteal cleft area. Scrotal infection alone is rare. It is more common in male
T. rubrum, *T. mentagrophytes* (found in 90 % cases)	Human	*Tinea unguium (onychomycosis)*: Onychomycosis is a common nail ailment associated with pain, discomfort and impaired/lost tactile functions. Toenail dystrophy can interfere with walking, exercise or proper shoe fit. In addition, it has both psychosocially and physically detrimental effects. Clinically four types of onychomycosis are observed such as distal and lateral subungual onychomycosis, proximal subungual onychomycosis, superficial white onychomycosis and total dystrophic onychomycosis
T. tonsurans	Human	*Tinea capitis*: It is the dermatophytic infection of hair/scalp. It is more common in children who remained in close contact with infected pet animals. The infection produces kerion formation which is characterised by a raised, tender mass of inflamed tissue
T. mentagrophytes, *T. tonsurans*, *T. rubrum*	Human	*Kerion celsi-type tinea capitis (tinea profunda capillitii)*: Deep-seated mycosis of scalp composed of tumorous mass with thick crusts
T. schoenleinii. *T. violaceum*, *T. verrucosum*, *zoophilic T. mentagrophytes* var. *quinckeanum*	Human	*Tinea capitis favosa*: It is a chronic inflammatory dermatophyte infection of the scalp and less commonly of the glabrous skin and nails. The classical lesion is concave, cup-shaped yellow crust on the scalp and glabrous skin that is associated with severe alopecia (scutulum)
T. tonsurans	Human	*Tinea gladiatorium*: It affects the upper part of the body and is common among wrestlers. It is transmitted by direct skin-to-skin contact
T. mentagrophytes, *T. verrucosum*, *T. rubrum*	Human	*Tinea barbiae (Barbers' itch)*: It is an uncommon dermatophyte infection of the beard and moustache areas. In previous years, it was more frequently observed due to use of common razor or blade in saloons. Clinically it shows sycosis, iododerma, contact dermatitis and perioral dermatitis
T. rubrum	Human	*Tinea incognito*: It results from lack of diagnosis and continuous corticosteroid application. The plaques with erythema and papules on the neck and breast area are noted
T. rubrum (Aspergillus and *Phoma*)	Human	*Majocchi's granuloma*: It is a folliculitic and perifolliculitic fungal infection of the dermis in both healthy and immunocompromised hosts. It presents as nodules, plaques and papules on areas that are prone to trauma. It generally appears on the upper and lower extremities (forearms, hands, legs, ankles) or on the scalp and face

4.1.13 Immunity

The innate and adaptive immune responses are generated against *Trichophyton*. The mammalian immune system has pattern recognition receptors (dectin-2, chi-lectins) which identify fungal pathogen-associated molecular patterns (PAMP) such as mannan and chitin. The mannan–dectin interaction can upregulate the pro-inflammatory cytokine (TNFα) production by macrophages. However, relationship between the chitin recognition and development of innate or adaptive immune response is not established. The antimicrobial peptide such as cathelicidin plays a role to limit the invasion of dermatophytes. During atopic dermatitis the level of cathelicidin is decreased which helps in invasion of dermatophytes. The fungal hyphae can activate complement via the alternative pathway and generate chemotactic factors for neutrophils. Further, during fungal invasion of keratin layer, the keratinocytes release IL8 which also attracts neutrophils in the stratum corneum. The neutrophils and monocytes can ingest and kill the spores of *T. rubrum*.

For development of adaptive immune response, the dermatophyte antigens are trapped and processed by the skin Langerhans' cells in the epidermis. The Langerhans' cells present the antigens in MHC class II restricted fashion to T cells in the lymph nodes draining the skin. The migration of Langerhans' cells for antigen presentation is initiated by the cytokine granulocyte-macrophage colony-stimulating factor (GM-CSF). In calves, experimentally infected with *T. verrucosum*, influx of different immune cells such as neutrophils, macrophages, CD4+ and CD8+ T lymphocytes, and γδ T cells were observed. The protective role of Th2-mediated immune response producing anti-*Trichophyton* antibodies in calves is questionable. In human, Th2-mediated response produces IL4 which induces antibody class switching to IgG4 and IgE, associated with immediate hypersensitivity (IH) reaction (wheal and flare reaction). The IH reaction is often detected in chronic infection which also indicates the inefficiency of humoural immunity in offering protection.

The activation of Th1 cells produces a cell-mediated immune response with the cytokine profile such as IFN γ, IL12 and IL2 which can recover the animals. This cell-mediated response is characterised by production of positive delayed-type hypersensitivity (DTH) reaction which occurs due to recruitment of cells into the skin associated with deposition of fibrin. It is also associated with lower titres of IgG directed towards *Trichophyton* antigens and the absence of IgE or IgG4. The efficient vaccine should activate the cell lineages which can produce an effective cell-mediated immune response to clear the infection.

The balance of Th1 and Th2 immunity determines the progress of *Trichophyton* infection, although, the dermatophytes prefer to induce Th2 response with B cell stimulation and inefficient macrophage activation for establishment of the infection.

4.1.14 Diagnosis

4.1.14.1 Clinical Specimen

The lesion area should be cleaned with 70 % ethanol to remove the bacterial contamination. The scales from the active lesions should be collected with a blunt scalpel (or toothbrush) and the affected hairs should be plucked from the lesion without breakage. For suspected onychomycosis, the nails should be cleaned with alcohol and affected nail bed should be exposed by removing the onycholytic nail plate with a nail clipper. The subungual hyperkeratotic material should be collected with a scalpel or curette. They are wrapped in brown paper (or coloured paper) and are kept in a tight container preferably without moisture for transport into the laboratory.

4.1.14.2 Laboratory Examination

1. *Direct examination*: The skin scrapings, hair including the hair bulb, nail fragments can be observed under microscope by wet mount with 10 % KOH preparation. The digestion of keratin with the KOH can be speed up with heat. An alternative preparation is 20 % KOH with 36 % DMSO that provides rapid

Table 4.5 Different species of *Trichophyton* associated with hair invasion

Ectothrix invasion	Endothrix invasion	Noninvasion
T. verrucosum	*T. schoenleinii*	*T. rubrum*
T. equinum	*T. violaceum*	*T. simii*
T. mentagrophytes	*T. tonsurans*	*T. concentricum*

diagnosis without heating and the specimens can be preserved for long time. The hyphae are observed under the microscope invading the hair and producing arthrospores. There are two types of hair invasion by the dermatophytes, known as ectothrix and endothrix invasion. In ectothrix invasion, the arthrospores are observed on the surface of the hair, whereas in endothrix invasion, the spores are found within the hair. There are some *Trichophyton* species also which cannot attack the hair. The different *Trichophyton* species associated with hair invasion are enlisted in Table 4.5.

The specimens can be stained with chlorazol black or Parkers blue ink. Chlorazol helps in identification of hyphae by staining the carbohydrate rich cell wall without staining the contaminants such as cotton or elastic fibres. The fine needle aspirates can be stained with the May–Grunwald–Giemsa method.

Majority of *Trichophyton* species (except *T. schoenleinii*) do not produce fluorescence under Wood's lamp. However, fluorescence microscopy (UVA, 365 nm) produces better resolution than wet mount with KOH. The fluorescent substance (calcofluor, acridinium orange or Blankophor) is added to the KOH. It binds fungal cell wall and increases the visibility of fungal hyphae and spores.

2. *Isolation and identification*: Media and incubation condition as described earlier will serve the purpose.
3. *Histopathology*: The histopathological slides are prepared with periodic acid–Schiff (PAS), Grocott methenamine silver and calcofluor white (CFW) stains. The histopathological section shows the dead and degenerating mycelium with cellular debris at the centre of the lesion with well preserved hyphae at the periphery of the lesion. In *T. schoenleinii* infection, the 'scutulum' (concave, cup-shaped yellow crust) is observed on the atrophic epidermis. Inflammation with round-cell infiltrate is seen in the adjacent dermis. In comparison to direct microscopy histopathological investigation is much reliable, although it cannot detect the species of dermatophyte. The diagnostic sensitivity can be increased with biopsy which is not always possible to conduct especially in human patients suffering with diabetes.
4. *Molecular biology*: The conventional PCR have developed for identification of different *Trichophyton* species using nuclear ribosomal DNA (rDNA), mitochondrial DNA (mtDNA), chitin synthase 1 (*chs1*), topoisomerase II (*TOP2*) genes, small (18S rRNA) and large subunit (28S rRNA) of ribosomal RNA as major target genes. Further, randomly amplified polymorphic DNA (RAPD), PCR fingerprinting, amplified fragment length polymorphism (AFLP) and sequencing of microsatellite markers, have been successfully applied to species identification in dermatophytes but were mostly unable to discriminate between the strains. The sequencing of the internal-transcribed spacer (ITS) region is commonly used for phylogenetic analysis and identification of dermatophyte species. Although, the ITS region shows high sequence similarities between the species of *Trichophyton* (90–100 %), the discriminating polymorphisms (barcode sequence) exists in ITS1 and ITS2 region of the members of *T. tonsurans* and *T. equinum*. In addition, non-transcribed spacer (NTS) region of the rRNA genes is also considered suitable for strain typing of some *Trichophyton* species, such as *T. rubrum*, *T. interdigitale* and *T. tonsurans*.

   Recent progress includes the use of PCR-ELISA which can specifically identify the PCR amplified product with the help of enzyme labelled probe producing colour reaction in positive cases. The uniplex-PCR-ELISA can identify *T. rubrum*, *T. interdigitale*, *T. tonsurans* and *T. violaceum* individually. The matrix-assisted laser desorption/ionisation time-of-flight mass spectrometry (MALDI-TOF-MS) can identify the species of *Trichophyton* from the grown cultures which is very fast and specific method and the result shows 98–99 % similarity with PCR.

4.1.15 Treatment

The most of the ringworm infection in animals is self-limiting; however, treatment is required for its zoonotic potential. The infected animals especially the pets and companion animals may transmit the infection to the animal handlers or pet owners who come in direct contact. The hairs of the affected part should be clipped properly before application of topical antifungals such as thiabendazole, miconazole, ecoconazole, ketoconazole, itraconazole, lime-sulphur solution, 5 % sodium hypochlorite solution, etc. If the topical application is ineffective, systemic antifungals such as ketoconazole, clotrimazole, itraconazole, terbinafine may be used. Among them, terbinafine is most effective. Griseofulvin is now a day avoided due to its acute toxicity.

In human, systemic treatment with antifungals is considered in tinea capitis, tinea unguium and during significant hyperkeratosis. In tinea capitis favosa, systemic treatment with terbinafine or itraconazole is preferred along with clearing away the scalp debris, removing the crusts and general improvement of scalp hygiene. In other cases, topical application of antifungals for 1–6 weeks is recommended.

The untreated patients of onychomycosis may act as reservoir and transmit the infection to other family members and neighbours during communal bathing (ponds). In diabetic patients, it carries the risk for development of diabetic foot. So, treatment of onychomycosis is required not only for cosmetic purpose but also for prevention of severe menace. When the infection covers more than 50 % of the distal nail plate and nail matrix, the systemic treatment with ketoconazole, itraconazole, fluconazole and terbinafine is recommended. The new therapies such as ciclopirox hydrolacquer based on water soluble biopolymer technology, 1 % fluconazole and 20 % urea in a mixture of ethanol and water and terbinafine nail solution are also promising. In more severe cases, surgical removal of the part (debridement) or whole (avulsion) of the nail plate may be followed.

Among plant-based antifungals, Neem (*Azadirachta indica* A. Juss) seed and leaf extracts showed antidermatophytic activity against *T. rubrum* and *T. violaceum* in vitro.

4.1.16 Vaccine

In several countries such as Russia, Hungary, former German Democratic Republic, former Yugoslavia, Norway and Bulgaria, the vaccine against *Trichophyton* is used since 1970. It is a non-adjuvinated, live vaccine containing freeze dried conidia and hyphal elements of *T. verrucosum* (LTF-130 strain). The vaccine is prophylactic as well as curative in nature which is inoculated intramuscularly at the side of the neck. In calves the recommended age for vaccination is 2–4 weeks. The immunity is life-long and no annual booster is required. If the vaccination is performed in the calves of 4–6 days old, a lesion (1–2 cm) appears at the injection site with hair loss and moderate scaling which regenerates after several weeks probably due to residual virulence of vaccine strain. The bovine ringworm is eradicated from Norway by vaccination and continuous monitoring.

The formalin inactivated vaccine against *T. verrucosum* is also produced which offered intermediate protection under field trial. Experimentally, formalin inactivated conidia and mycelia elements of *T. equinum* along with adjuvant offered 75 and 87 % protection in horses, respectively.

4.2 *Microsporum*

David Gruby (1843) described tinea capitis (favosa) in human and isolated *Microsporum* as causative agent with fulfilment of Koch's postulate. He coined the genus name *Microsporum* on the basis of small spore production around the hair shaft. Gruby also first described the type species *Microsporum audouinii* in the name of his close associate Victor Audouin, the then Director of the Museum of Natural History in Paris. In 1881, Megnin first described *M. gallinae*. White (1899) reported 279 cases of human ringworm in Boston during the period

October 1895 to July 1898. Of the 279 cases, 139 (50 %) were caused by *Microsporum audouinii* in children. The existence of the sexual phase of certain dermatophytes was first recognised by Nannizzi (1927) in Italy. He also described the sexual form of *Microsporum gypseum* as *Gymnoascus gypseum*. Griffin validated the Nannizzi's investigations and confirmed the existence of the sexual phase of *Microsporum*. This sexual stage was renamed as *Nannizzia* in the name of Nannizzi. In1934, Emmons first classified the dermatophytes into three genera, i.e. *Trichophyton*, *Microsporum* and *Epidermophyton*. This description is till date considered as the standard system of classification for dermatophytes. Bodin first described two species of *Microsporum* such as *M. canis* and *M. gypseum*. Kligman (1952) described the epidemics of tinea capitis due to *Microsporum audouinii* which was known as the Atlantic City board epidemic among the school children.

In India, Dey and Kakoti (1955) first reported the isolation of *Microsporum gypseum* from ringworm lesion in a laboratory rabbit. Kandhari and Sethi (1964) established the presence of *M. canis* infection in dogs in Delhi. The presence of *M. nanum* in the Indian pigs was reported by Gupta et al. (1968), whereas *M. gypseum* infection in other domestic animals such as cattle (Gupta et al. 1970), sheep (Thakur et al. 1983), goat (Thakur and Verma 1984), horse (Tewari 1962; Gupta et al. 1970), pets (Pal et al. 1990) and wild animal (Pal 1988) was also reported from India.

4.2.1 Morphology

The septate, branching hyphae are produced by *Microsporum* in the artificial culture and nonparasitic (environmental) state. Like other dermatophytes, *Microsporum* also produce asexual spores known as conidia. Two types of conidia, i.e. macroconidia (macroaleuriospores) and microconidia (microaleuriospores), are detected. The size of macroconidia varies from 6 to 160 μm × 6–25 μm. The macroconidia are rough walled, spindle shaped, fusiform (*M. vanbreuseghemii*) or ovoid (*M. nanum*). The wall of macroconidia is thin, moderately thick or thick in nature which can traverse inside to produce 1–15 septa. They are generally arranged singly along the hyphae (Figs. 4.3 and 4.4). The microconidia are sessile or stalked, less abundant, clavate in shape and arranged singly along the hyphae.

4.2.2 Classification

All dermatophytes belonged to the phylum Ascomycota, class Euascomycetes, order

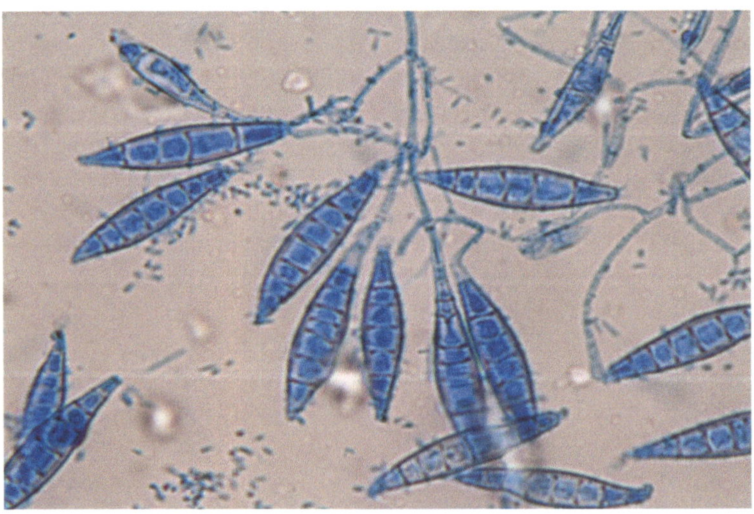

Fig. 4.3 Macroconidia of *Microsporum canis* (Photograph courtesy – Prof. Alexandro Bonifaz, Head, Department of Mycology and Dermatology service, General Hospital of Mexico, Mexico City)

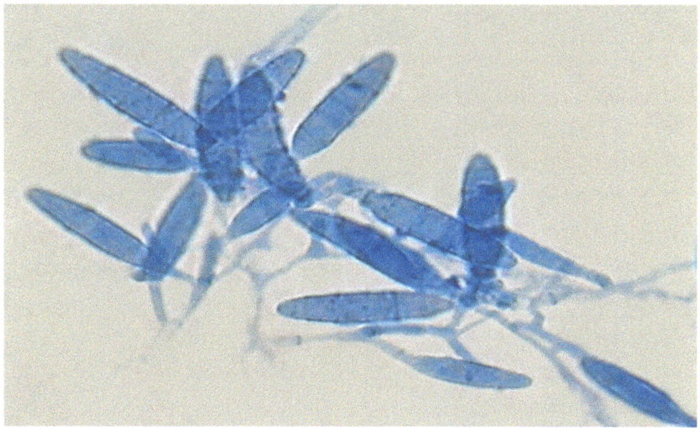

Fig. 4.4 Macroconidia of *Microsporum gypseum* (Photograph courtesy – Prof. Alexandro Bonifaz, Head, Department of Mycology and Dermatology service, General Hospital of Mexico, Mexico City)

Table 4.6 Ecological niche-based classification and host specificity of *Microsporum*

	Zoophilic (animal associated)			
Anthropophilic (human associated)	Fungi	Major host	Isolated from other animals	Geophilic (soil associated)
M. audouinii (isolated from dogs also)	*M. canis*	Dog, cat	Cattle, sheep, horse	*M. nanum* (isolated from pigs)
M. ferrugineum	*M. equinum*	Horses	–	*M. gypseum* (isolated from dogs, cats)
	M. gallinae	Fowl	Cattle, dog, cat	*M. amazonicum*
	M. persicolor	Vole	Dog	*M. boullardii*
				M. cookei
				M. praecox
				M. racemosum
				M. ripariae
				M. vanbreuseghemii
				M. duboisii

Onygenales and family Arthrodermataceae. The family Arthrodermataceae contains four morphological anamorph genera such as *Trichophyton*, *Microsporum*, *Epidermophyton* and *Chrysosporium*. The genus *Microsporum* contains 18 numbers of species. The type of species is *Microsporum audouinii*. The species complexes are *Arthroderma otae* complex which consists *Microsporum canis* (including the old taxa *M. equinum*, *M. distortum*), *M. audouinii* (including var. *rivalieri* and *langeronii*) and *M. ferrugineum*, and *Microsporum gypseum-fulvum* complex which consists *Arthroderma incurvata*, *A. gypsea* and *A. fulva*. Clinically important species are *M. canis*, *M. gypseum* (gardener ringworm), *M. gallinae*, *M. nanum* and *M. audouinii*.

The ecological niche-based classification system describes *Microsporum* as anthropophilic (human associated), zoophilic (animal associated) and geophilic (soil associated) fungi like other dermatophytes. The examples of different *Microsporum* species are enlisted in Table 4.6.

4.2.3 Reproduction

The sexual reproduction is observed in geophilic and zoophilic species of *Microsporum*. The mating competent species have an asexual type (anamorphs) and a sexual type (teleomorph). The sexual stage (teleomorph) of *Microsporum* and *Trichophyton* was previously known as *Nannizia* and *Arthroderma*, respectively. Due to similarities in the properties of both genera, *Nannizia* is replaced with *Arthroderma* at present. The teleomorphs of different *Microsporum* species are described in Table 4.7.

Table 4.7 Teleomorphs of *Microsporum* species

Microsporum species (anamorph)	Teleomorph
M. canis var. *canis*, *M. canis* var. *distortum*	*Arthroderma otae*
M. nanum	*A. obtusum*
M. gypseum	*A. gypseum*
M. gypseum	*A. incurvata*
M. vanbreuseghemii	*A. grubyi*
M. fulvum	*A. fulvum*
M. persicolor	*A. persicolor*
M. racemosum	*A. racemosum*
M. amazonicum	*A. borelli*
M. cookei	*A. cajetani*
M. boullardii	*A. corniculata*
Unnamed	*A. cookiella*

The sexual reproduction of other ascomycetous fungi such as *Aspergillus* is controlled by *MAT* locus with two idiomorphs such as *MAT-1* and *MAT-2*. The heterothallic *MAT-1* and *MAT-2* can undergo sexual mating. The similar kind of *MAT* locus is identified in two species of *Microsporum* such as *Microsporum gypseum* (geophilic) and *Microsporum canis* (zoophilic) which indicates their mating competence. However, the teleomorph of *Microsporum* (*Arthroderma*) isolates always exhibit only one mating type which shows the possibility of interspecies sexual recombination as observed in mating between *T. rubrum* and *Arthroderma simii*.

4.2.4 Susceptibility to Disinfectants

The spores of *Microsporum* are susceptible to lime sulphur (1:33), 0.2 % enilconazole and chlorine bleach (1:10 to 1:100)

4.2.5 Natural Habitat and Distribution

The anthropophilic and zoophilic *Microsporum* primarily inhabits human and animals, respectively, although they can infect vice versa. The geophilic *Microsporum* prefers to utilise the keratinous materials (hair, feathers, hooves, horns) after dissociation from living animals and which are in the process of decomposition in the soil. Some of the geophilic *Microsporum* (*M. gypseum*) can survive in the environment in absence of keratinous materials also such as in sewage sludge, sediment/soil within the cave and cave wall. The infectious agent of *Microsporum* (anthropophilic) is chlamydospores or arthrospores which can survive in the environment for several years and are heat resistant especially within the skin scales.

Microsporum is worldwide in distribution, although some species are endemic in limited geographical location. *M. canis* is the major etiological agent of tinea capitis and tinea corporis in some parts of Europe (Slovenia, Croatia and Italy), the eastern Mediterranean, South America, Asia (Iran, Yemen and Palestine) and Africa (Libya). Whereas *M. audouinii*, causing tinea capitis in children, was restricted within North America but currently is distributed in Asia and Africa. In Germany, it was considered as major etiological agent of favus in humans. After introduction of griseofulvin (1958), they are almost eradicated from Germany and other parts of Central Europe. However, *M. audouinii* is still prevalent in African countries (Malawi, Western Kenya), causing tinea capitis in human. *M. gypseum* was isolated from human dermatophytic lesions in Asia (in Iran it shares 4.1 % of the dermatophytic lesions) and Africa (Nigeria). *Microsporum ferrugineum* is now restricted to a few rural areas of Asia and Africa.

M. canis is the predominant species causing dermatophytosis in dogs and cats in Italy which is followed by geophilic species such as *M. gypseum*. In Iran, stray cats were reported to harbour *M. canis* without showing any symptom. In Southern India (Chennai), prevalence of *M. gypseum* complex was detected among the stray dogs.

4.2.6 Genome

Currently whole genome sequences of two *Microsporum* species (*M. canis* and *M. gypseum*) are available at Broad Institute's dermatophyte comparative database (http://www.broadinstitute.org). The genome statistics of *M. canis* and *M. gypseum* is described in Table 4.8.

Table 4.8 Genome statistics

Genome properties	*M. canis*	*M. gypseum*
Size (Mb)	23.1	23.2
GC (%)	47.5	48.5
Number of protein-encoding genes	8,915	8,907
Mean coding sequence length (nt)	1,459	1,436
Mean exon no. per gene	3.21	3.15
Number of tRNAs	82	83

The mitochondrial DNA (mtDNA) of *M. canis* and *M. nanum* is circular in nature and 23,943 and 24,105 bp in size, respectively, which contains 15 protein-encoding genes. The mitochondrial genome size of *M. canis* is the smallest genome of the studied dermatophytes. The GC value of mitochondrial genome is 24 % and it contains 2 rRNAs, 25 tRNAs and 1 intron. The coding region genes are highly similar (>90 %) between the species, whereas the differences between species exist in the introns and intergenic regions.

4.2.7 Isolation, Growth and Colony Characteristics

The conventional solid medium includes Sabouraud dextrose agar with cycloheximide, chloramphenicol and gentamicin. The cultures are incubated at 25–30 °C for 2 weeks. The dermatophyte test medium (DTM) is not suitable for *Microsporum* isolation, as some of the species produce false-negative result in this medium. For demonstration of macroconidia, potato dextrose agar and other sporulation media are used.

The primary colonies are large, regular or irregular spreading types with varying pigmentation. For example, *M. canis* produces lemon yellow pigment which may spread throughout the medium and *M. nanum* produces rusty colonies. The colonies later become velvety (*M. audouinii*) or powdery (*M. gypseum*) due to production of white fluffy tufts of aerial mycelium on the surface of colonies which results in the loss of pigmentation and conidiation. It is known as 'pleomorphism of dermatophytes' which results from single chromosomal gene mutation.

4.2.8 Antigenic Characteristics

The ribosomal fraction of *M. canis* showed certain degree of immunodominance when used as experimental vaccine in guinea pigs. However, these antigens were not characterised properly. *Microsporum* produces two types of proteases like other dermatophytes such as serine protease (Sub3) and zinc metalloprotease (Mep3). The recombinant *M. canis* sub3 and Mep3 can induce cell-mediated immune responses in experimental guinea pigs. The antibody response was observed against Mep3, not against Sub3. Experimental vaccination in guinea pigs with the proteases did not offer any protection.

Recent studies showed the presence of nine IgG-binding proteins (48–180 Kda) as immunodominant antigens in the sera of *M. canis*- and *M. gypseum*-infected dogs.

4.2.9 Virulence Factors

The virulence factors of *Microsporum* explored by different investigations till date are enlisted in Table 4.9.

4.2.10 Transmission

Transmission of *M. canis* arthrospores in animals occur through direct contact with the infected or carrier animals especially the cats and dogs. The hair fragments containing the spores are major source of infection. The risk factors include introduction of new animals in house/shed, animal shows/market, mating, etc., where the animals come in close contact with each other. Although indirectly, the spores are transmitted through brushes, collars and grooming utensils. Sometimes the animals get infected with *M. gypseum* (geophilic) during their burrowing or playing with contaminated soil.

The direct contact with infected cats, dogs and rabbits may transmit the *M. canis* infection in human. Human-to-human transmission is rare

Table 4.9 Virulence mechanisms and factors possessed by *Microsporum*

Virulence factors	Functions
Endoprotease ['Ekase' (keratinase) which is composed of two following enzymes]	They help to degrade the keratin
(a) Subtilisins (Sub3, serine proteases, 34 Kda, pH 7–9 is required for optimum activity)	
(b) Fungalysins (Mep3, 48 Kda, Zn metalloproteases with HEXXE motif. Their optimum pH of activity is between 7.0 and 8.0.)	
Pseudokinase (inactive kinase)	They help in the adaptation of the fungi in diverse animal hosts by increasing the tolerance against environmental challenges. Specifically they compete for the substrates with active kinase and act as inhibitory pseudo-substrates
Exoproteases	The Dpp can degrade the large peptides into oligopeptides and free amino acids
(a) Dipeptidyl peptidases (DppIV, DppV)	
(b) Carboxypeptidases	
Sulphite	In the presence of sulphite, disulphide bonds of the keratin substrate are directly cleaved to cysteine and S-sulphocysteine. The reduced proteins become accessible to other proteases for further cleavage
LysM along with polysaccharide deacetylases (type I)	(a) Probably LysM protein conceals the dermatophyte cell wall components and carbohydrates and thus avoids the host immune response to the fungi (b) The polysaccharide deacetylase helps in chitin catabolism and is involved in chitin scavenging in the soil, defence from other fungi or cell wall modification
Polycyclic aromatic prenyltransferases (pcPTases)	It causes prenylation of aromatic compounds which can increase lipophilicity of the compounds and enhances their interactions with biological membranes and proteins

although observed in neonatal and intensive care units where an infected nurse was detected as the source of infection. Among children no gender predilection is observed, whereas in adults, women are infected more frequently than men, with a ratio ranging from 3:1 to 6:1.

4.2.11 Pathogenesis

Microsporum spores cannot invade the healthy skin of the animals after transmission. So many animals such as cats and dogs become carrier of *M. canis* arthrospores. This carrier stage may progress to infection which depends on certain predisposing factors such as young age, immunosuppression (virus or drug induced), nutritional deficiency (protein, vitamin A) and high environmental temperature with high humidity and skin trauma (injury or scratches due to clipping, playing, aggressive behaviour, ectoparasite infestation, pruritus). After penetration through the skin trauma, the *Microsporum* spores germinate in the stratum corneum. The hyphae grow along the hair shaft to the follicles where they produce spores. The spores remain attached by forming a thick layer around the hair shaft. As *Microsporum* cannot colonise the deeper part of the skin or hair follicles, the hairs grow normally but break near the skin surface causing alopecia. The fungal metabolites induce inflammatory reaction at the site of infection. The inflammation causes the pathogen to move away from the site of infection in a circular way with healing at the centre and papules at the periphery which is responsible for classical ringed lesion.

The invasion of stratum corneum may occur in immunosuppressed animals with the help of the virulence factors (Table 4.9). It produces generalised skin illness with secondary bacterial infection. Rarely, a marked inflammatory response against the hyphae produces a nodular granulomatous reaction in dermis (pseudomycetoma) in Persian cats.

In immunocompetent animals which are kept in hygienic conditions, the *Microsporum*-induced ringworm lesions are limited in a restricted area (head in cats) which disappears after several weeks or after proper treatment.

4.2.12 Disease Produced

The major animal and human diseases produced by *Microsporum* are enlisted in Table 4.10.

Table 4.10 Major diseases of animal and human caused by *Microsporum*

Fungi	Host	Disease
M. canis, *M. gypseum*	Cattle	*Bovine ringworm*: The calves are most susceptible especially when kept within the crowded pens. The lesions are circular, painless, thick, white and scattered. The lesions appear in the head, neck, paw and less frequently in the back, flank and limbs. The secondary bacterial infection may occur associated with pruritis. Other clinical signs are skin scaling and loss of hair
M. gypseum	Horses	*Equine dermatophytosis*: The dry, scaly and multiple lesions appear in any part of the body especially in the groomed part. Sometimes the lesions become larger to produce a 'moth-eaten appearance'
M. gallinae	Poultry	*Favus (white comb)*: Birds of all age groups kept under poor management are susceptible. White patches are noticed in the comb of male birds. The lesions may coalesce and the whole comb is covered with thick white coating. The infection may progress into the feather, gastrointestinal tract and respiratory tract. Necrotic foci with yellowish mass are observed in trachea. Emaciation followed by death may occur
M. nanum, *M. canis*, *M. gypseum*	Pigs	Large breeds (Yorkshire) and all age groups are equally susceptible. Unhygienic condition, high humidity, high stocking density favours the transmission. The lesions are small, circular, mildly inflamed and are located throughout the body. The skin becomes reddish with formation of crusts. No alopecia is observed. The disease is self-limiting and resolves within 2–3 months
M. canis	Cats	*Feline ringworm*: The infection causes regular and circular alopecia (3–6 cm in diameter) with desquamation, erythema in the margin and central healing. Lesions are single or multiple, mostly confined in the head. Occasionally lesions are found in legs and tails. Intensity of pruritis is variable and usually no fever or anorexia is observed. 'Miliary dermatophytosis' affects dorsal trunk. In immunosuppressed patients, larger area of alopecia, erythema, pruritis and crusts develops. Rarely nodular granulomatous dermatitis (pseudomycetoma) develops. Feline otitis is also produced which is characterised by waxy, ceruminous discharges from the ear canal
M. canis, *M. gypseum*	Dogs	The lesions are circular (2.5 cm in diameter) and are located in face, elbows, paws. The skin becomes dry and scaly
M. audouinii, *M. canis*	Human	*Tinea capitis*: It is dermatophyte infection of the scalp and hair shaft, commonly occurs in children. The transmission occurs due to close contact with infected pets. The infection causes mild erythema, patchy area of scaling, grey hair stumps. Sometimes the infection causes huge inflammation, folliculitis and kerion (boggy, sterile, inflammatory scalp mass) formation, areas of scaring and alopecia. The swollen hairs fracture a few millimetres from the scalp and produce 'black dot' alopecia. *M. audouinii* causes ectothrix type of hair invasion
M. canis, *M. gypseum*	Human	*Kerion celsi-type tinea capitis (tinea profunda capillitii)*: Deep-seated mycosis of scalp composed of tumorous mass with thick crusts
M. canis	Human	*Onychomycosis*: The clinical manifestation involves distal subungual onychomycosis, onychodystrophy, proximal subungual onychomycosis. *M. canis* generally does not invade the nails
M. canis	Human	*Tinea incognito*: It results from lack of diagnosis and continuous corticosteroid application. The plaques with erythema and papules on the neck and face with hair loss are noted
M. canis	Human	In South-Asian countries (Thailand), women suffer from chronic respiratory disease which is associated with lung cancer

4.2.13 Immunity

Innate immunity is the first line of defence against *Microsporum* infection in animals or human. Dectin 2 acts as pattern recognition receptor (PRR) which can bind the hyphal fragments of *Microsporum*. The pro-inflammatory cytokines are required for activation of immune cells to produce an effective immunity. *M. canis* can induce IL1β (pro-inflammatory cytokine) secretion through the activation of inflammasome (intracellular multiprotein complex, NLRP3). The activation of NLRP3 occurs through the sequential stimulation of cathepsin B, K (+) efflux and reactive oxygen species.

The pathways and the molecules involved in production of adaptive immunity (Th1 and Th2 mediated) against *Microsporum* are not specifically elucidated, although, the similarities in pathways with anti-*Trichophyton* immunity are assumed. The Th1 response (cell-mediated immunity) is associated with delayed-type hypersensitivity (DTH) and recovery from *Microsporum* infection. The study indicated the presence of larger area in DTH reaction in the cats recovered from *M. canis* infection than the cats still having infection and which were never exposed. The antibodies produced in Th2 immunity (humoural) can act as fungistatic due to opsonisation and complement activation properties. Their exact role in recovery and protection is unexplored.

4.2.14 Diagnosis

4.2.14.1 Clinical Specimens

The hairs and scales can be collected from the margin of the lesion after gentle swabbing with alcohol for decontamination of the skin. In subclinically infected animals, brushing the suspected skin over the back, shoulders, sides, hindquarters and legs with an unused toothbrush or hair brush can be performed. The brushes are transported to the laboratory in brown paper or plastic bags. The hairs and scales are also wrapped in brown paper (or coloured paper) and are kept in a tight container, preferably without moisture for transport into the laboratory. The brushes should be processed on the same day of collection.

For geophilic *Microsporum* (*M. nanum*, *M. gypseum*, *M. cookie*), the soil surface sampling is preferred. The top layer is removed and the representative samples are scooped from the lower layer of the soil. The soil should be processed on the day of collection, if stored, 0–4 °C is adequate.

4.2.14.2 Laboratory Examination

1. *Direct examination*:
   The process of direct examination is same as described earlier (Sect. 4.1).
   Most of the *Microsporum* species produces *ectothrix* type of hair invasion. For example, ectothrix invaders are *M. canis*, *M. equinum*, *M. audouinii*, *M. gypseum* and *M. nanum*.
   Majority of *M. canis* strains (>50 %) produces green fluorescence due to tryptophan metabolite under Wood's lamp ($\lambda =$ 366 nm). Sometimes medication, secretions and artificial fibres may hinder the fluorescence.
2. *Isolation and identification*: Media and incubation condition as described earlier will serve the purpose.
3. *Molecular biology*: A multiplex PCR is developed which can simultaneously detect *Trichophyton* spp., *Microsporum canis* and *M. audouinii* from the clinical specimens. The RAPD-PCR was developed for differentiation of *M. canis* strains isolated from different species such as dogs, cats, human, fox and rabbit. Sometimes the band pattern generated by this molecular technique produces ambiguous results which can be resolved by genomic in situ hybridisation (GISH). Using specific GISH probes, discrimination between *Trichophyton interdigitale*, *Trichophyton rubrum* and *Microsporum canis* has been conducted. The uniplex PCR-ELISA can also individually identify *M. canis*.

4.2.15 Treatment

In cats, the mild lesions recover spontaneously, although the treatment can reduce the course of the disease and transmission risk into the human. Topical application of antifungals is not so effective due to undetected exact position of lesions, poor penetration of drugs through the thick hair coat and lack of tolerance in some breeds. The systemic treatment should be followed at least for 10 weeks, preferably up to when the lesions heal and become culture negative. The antifungals used in systemic treatment of cats include itraconazole (better tolerance with less reported toxicity except anorexia), terbinafine (side effects include vomition, pruritus), ketoconazole (major adverse effects are liver toxicity, vomition, diarrhoea, anorexia, contraindicated in pregnancy) and griseofulvin (anorexia, vomition, diarrhoea, bone marrow suppression especially in Siamese, Himalayan, Abyssinian cats). The drug of choice is itraconazole which is administered by pulse therapy. The pulse therapy includes administration of ketoconazole at 5 mg/kg body weight/day for 1 week in every 2 weeks which should be continued for 6 weeks. Use of *M. canis*-based vaccine produces poor immunity against the pathogen in cats.

In human, treatment of tinea capitis duc to *M. canis* infection is difficult than the infection with *Trichophyton* probably due to ectothrix type of hair invasion and small size of the spores which make them inaccessible to the antifungals. Commonly used oral antifungals against *M. canis* are griseofulvin (microlyophilised form) and terbinafine. The fungistatic itraconazole can be used in children to treat kerion celsi type of tinea capitis along with topical antifungals and antibacterials, even with the poor response of *M. canis* to topical antifungals.

Among plant-based antifungals, Neem (*Azadirachta indica* A. Juss) seed and leaf extracts, *Curtisia dentata* (triterpenoids) extracts showed antidermatophytic activity against *M. nanum* and *M. canis*, respectively, in vitro.

4.3 *Epidermophyton*

The earlier report of *Epidermophyton* in the scientific literature was detected in 1870 when Harz described tinea cruris due to *Acrothecium floccosum*. Sabouraud (1910) classified the causal agents of dermatophytosis into four groups, i.e. *Achorion*, *Epidermophyton*, *Trichophyton* and *Microsporum*. In 1923, Ota and Langeron finally proposed the nomenclature of *Epidermophyton floccosum* which was originally described by Harz (1870). Emmons (1934) modernised the classification system and divided all dermatophyte species into three genera, i.e. *Microsporum*, *Trichophyton* and *Epidermophyton*, and declared that the genus *Achorion* is unnecessary.

4.3.1 Morphology

Epidermophyton produces septate hyphae like other dermatophytes with production of macroconidia. No microconidia are produced. The macroconidia are smooth, thick or thin walled with 1–9 septa, club shaped (clavate) and occur singly or in a cluster of 2–6 cells (Fig. 4.5).

4.3.2 Classification

All dermatophytes belonged to the phylum Ascomycota, class Euascomycetes, order Onygenales and family Arthrodermataceae. The family Arthrodermataceae contains four morphological anamorph genera such as *Trichophyton*, *Microsporum*, *Epidermophyton* and *Chrysosporium*. The genus *Epidermophyton* possesses 2 species which are *E. floccosum* and *E. stockdaleae*. The type species and only pathogenic species under the genus is *E. floccosum*. The ecological niche-based classification system describes *E. floccosum* as anthropophilic.

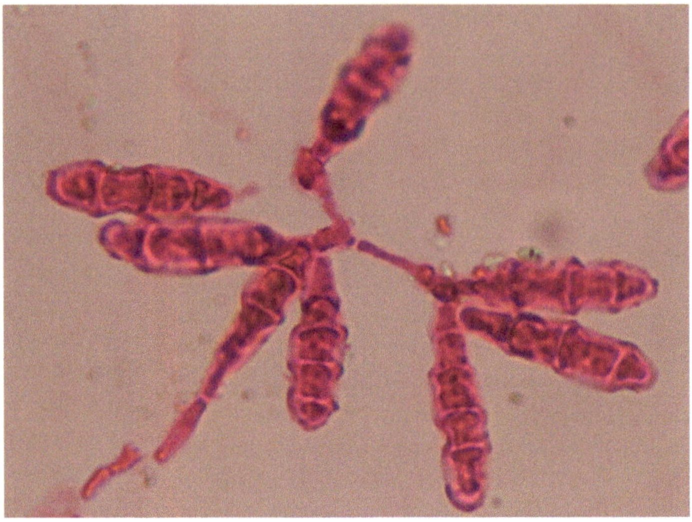

Fig. 4.5 Macroconidia of *Epidermophyton floccosum* (Photograph courtesy – Prof. Alexandro Bonifaz, Head, Department of Mycology and Dermatology service, General Hospital of Mexico, Mexico City)

4.3.3 Susceptibility to Disinfectants

The dermatophyte spores are susceptible to benzalkonium chloride, 1 % sodium hypochlorite, enilconazole (0.2 %), formaldehyde, iodophors, glutaraldehyde, phenolic compounds, benzylammonium bromide and ethoxyllauric alcohol. The last two are especially effective against the anthropophilic dermatophyte such as *E. floccosum*.

Recent finding indicated that *Candida albicans* releases volatile compounds such as dihydrofarnesol (R-DHF) and (2E, 6E) farnesol (F-ol) which can prevent the growth of dermatophytes such as *T. rubrum*, *T. mentagrophytes*, *M. canis* and *E. floccosum* in a concentration dependent manner. *E. floccosum* is completely destroyed by 12.5 mcg/mL dihydrofarnesol.

4.3.4 Natural Habitat and Distribution

E. floccosum is associated with human due to its anthropophilic nature. Rarely it is isolated from dogs or other animals. The other species (*E. stockdaleae*) is soil associated and nonpathogenic. The arthrospores (asexual spore) of *E. floccosum* produced within the human body may survive in the environment for prolonged period.

It is worldwide in distribution.

4.3.5 Genome

The information regarding the whole genome of *E. floccosum* is currently not available. Although the complete sequence of the 30.9 kb mitochondrial (mt) genome is reported, the major genes present in the mitochondrial genome include reduced nicotinamide adenine dinucleotide-ubiquinone oxidoreductase (nad1, nad2, nad3, nad4, nad4L, nad5, nad6), cytochrome oxidase (cox1, cox2, cox3), apocytochrome b (cob), ATP synthase (atp6, atp8, atp9), the small and large ribosomal RNAs (rns and rnl) and 25 tRNAs. The order of the genes present in the mtDNA of *E. floccosum* is similar with *Trichophyton rubrum* mtDNA with the exception of some tRNA genes. The phylogenetic analysis confirms *T. rubrum* as a close relative of *E. floccosum*.

4.3.6 Isolation, Growth and Colony Characteristics

No specific growth requirement of *Epidermophyton* is detected. The conventional solid medium includes Sabouraud dextrose agar with cycloheximide and chloramphenicol. The dermatophyte test medium (DTM) is also suitable for their isolation. The cultures are incubated at 25–30 °C for 2 weeks. Some authors suggested

30 °C as an ideal temperature for growth of *Epidermophyton*.

The colonies of *E. floccosum* in Sabouraud dextrose agar are velvety, greenish-brown or khaki coloured with orange-brown reverse pigmentation, radial grooves and folded centre. The older cultures show white-coloured tufts of mycelium.

4.3.7 Antigenic Characteristics

The study indicated the presence of 31 proteins in SDS-PAGE profile of *E. floccosum* ranging between 12.5 and 97.5 KDa. Among them, 20.1 and 18.4 KDa proteins are identified as immunodominant antigens.

No common antigens have been demonstrated between *T. mentagrophytes*, *M. canis* and *E. floccosum*.

4.3.8 Virulence Factors

E. floccosum are able to grow at 37 °C and can produce proteinase enzymes in vitro which are not well characterised. However, the gene encoding zinc metalloprotease (Mep1) is reported to be present in the isolates.

4.3.9 Transmission

The anthropophilic dermatophyte such as *E. floccosum* produces arthrospores within the hyphae in a vertebrate host (human or rarely in animals) and macroconidia outside the hyphae in the environment or artificial culture. The viability of the spores in the environment is longer than the other dermatophytes. *E. floccosum* is transmitted by these spores between the hosts or from environment into the host. The fomites such as brushes and clippers, and the common items used in showers and gymnasiums play major role in transmission among human.

4.3.10 Pathogenesis

After transmission of the spores, *E. floccosum* prefers to germinate at the stratum corneum of the skin without further penetration in the immunocompetent hosts. The sporadic invasion is reported in immunocompromised patients (Behcet's syndrome). Further, *E. floccosum* is not known to invade hair also. However, surface erosion of hair cortex and rare report of hair parasitism was observed.

4.3.11 Disease Produced

E. floccosum causes tinea corporis, tinea pedis, tinea unguium, onychomycosis and tinea cruris (occasionally) affecting inguinal areas particularly in males.

E. floccosum causes rare infection in animals. The sporadic reports of its isolation from dermatophytic lesions of dogs with or without immunosuppression (chronic hyperadrenocorticism) and other animals are available.

4.3.12 Diagnosis

4.3.12.1 Clinical Specimens

The lesion area should be cleaned with 70 % ethanol to remove the bacterial contamination. The scales from the active lesions should be collected with a blunt scalpel (or unused toothbrush). For suspected onychomycosis, the nails should be cleaned with alcohol and affected nail bed should be exposed by removing the onycholytic nail plate with a nail clipper. They are wrapped in sterile brown paper (or coloured paper) and are kept in a tight container preferably without moisture for transport into the laboratory. The loam soil is the most useful sample for isolation of soil associated *Epidermophyton*.

4.3.12.2 Laboratory Examination

1. *Direct examination*: The skin scrapings can be observed under microscope by wet mount with 10 % KOH preparation as described

earlier (Sects. 4.1 and 4.2). As *E. floccosum* never invades hair in vivo, no fluorescence is detected in wood's lamp examination.

2. *Isolation and identification*: Media and incubation condition as described earlier will serve the purpose.
3. *Molecular biology*: The conventional diagnostic PCR is employed targeting the DNA topoisomerase II gene for detection of *E. floccosum* from cultures. Further, the uniplex PCR-ELISA can also specifically identify *E. floccosum* directly from clinical specimens. The dermatophyte-specific single tube real-time PCR assay is also developed which can identify the individual dermatophyte species including *E. floccosum* directly from clinical samples such as nails.

4.3.13 Treatment

In tinea pedis and tinea cruris, topical application of terbinafine cream once daily for 1 week and butenafine 1 % cream applied once daily for 2 weeks can produce effective result. Systemic (oral) antifungal drugs (as described in Sects. 4.1 and 4.2) may be necessary in severe cases, or if the infection does not respond to treatment or reappears.

The leaves and stem powder prepared from *Commelina cyanea* (indigenous plant common in Northwest Cameroon) and ethanolic extract of propolis (wax-like substance) were detected to have in vitro antifungal activity against *E. floccosum*.

4.4 *Aspergillus*

Aspergillus (*L. aspergillus* – a special type of brush) is a filamentous, most widespread fungal genus containing both the pathogenic and beneficial species producing antibiotics, antifungals and antitumour drugs. In 1729, Micheli first described *Aspergillus* who found the similarity between the spore chain of the fungi with the brush ('Aspergillum') used for sprinkling holy water in the churches. In 1749, Reaumur first described avian aspergillosis in birds which was followed by report of similar syndrome in duck (Montagu 1813). The pathogenic *Aspergillus* (*A. candidus*) was detected in air sac lesion of a bullfinch in 1842 by Rayer and Montagene. Whereas *A. fumigatus* was first detected in the lung of a great bustard (*Otis tarda*) in 1863 by Fresenius, who was the first to use the term 'aspergillosis' for this respiratory disease. The major members of aflatoxins were first detected and its association with *Aspergillus flavus* was established in 1961 during investigation of mysterious 'turkey-X disease', causing high mortality in turkey poults (Sargeant et al. 1961). In 1962, the name 'aflatoxin', using first letter from 'Aspergillus' and the first 3 letters from 'flavus', was proposed.

In India, Datta (1938) first reported the association of *Aspergillus* with bovine haematuria affecting the kidney and bladder. Asthana (1944) first isolated *Aspergillus* from the lungs, caeca, intestine, kidney, ovary and testes from poultry in India. Pal (1996) first reported guttural pouch mycosis in horses.

4.4.1 Morphology

The mycelium is composed of septate hyphae (3–6 μm) with dichotomous branching. The primary branch originating from the vegetative hypha is known as *foot cell* which is further branched into the *conidiophores* (300 μm in length and 5–8 μm in diameter). The condiophores contain flask-shaped vesicles, 20–30 μm in diameter. The distal half of the vesicle is covered with a single series of *phialide*, 6–8 μm in length, arranged upwards paralleling the axis of the conidiophore. From these phialides, the chains of conidia (asexual spores) originate (Fig. 4.6). The conidial chain length occurs up to 400 μm (*A. fumigatus*). In the tissues, only mycelium is observed, and in the body cavities filled with air (air sac, nasal passage), the conidiophores with the phialides are found. Different species of *Aspergillus* can be identified on

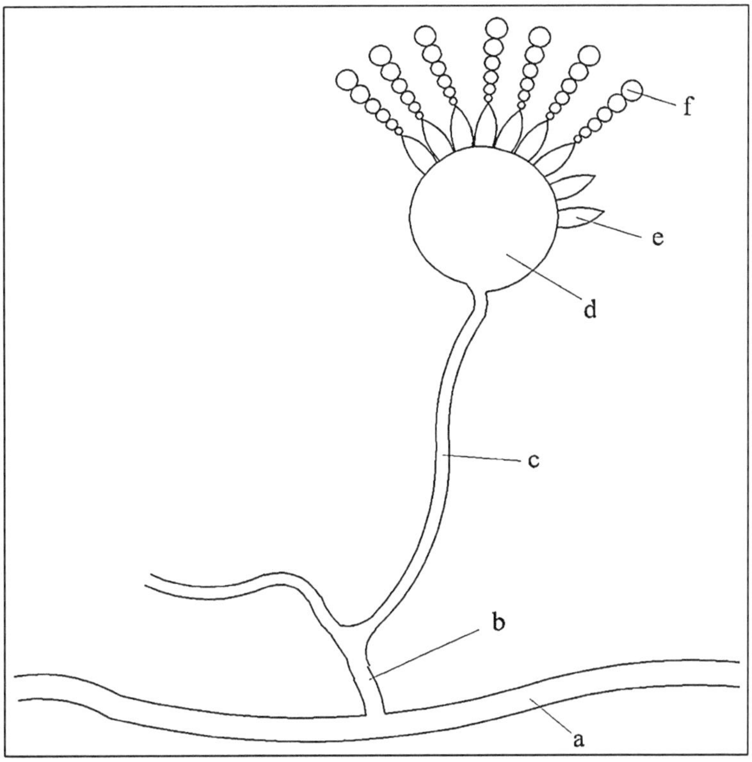

Fig. 4.6 Conidial arrangement of *Aspergillus* (schematic diagram). *a* Hypha, *b* footcell, *c* conidiophores, *d* vesicle, *e* phialide, *f* conidia

the basis of their microscopic morphological appearances.

The cell wall of *Aspergillus* (*A. flavus*) consists of glycoproteins, β-(1, 3) glucan, β-(1, 6) glucan, galactomannan and chitin. In *A. fumigatus*, the cell wall is composed of β-(1, 3) glucan, β-(1, 3/1,4) glucan, β-(1, 6) glucan, α (1, 3) glucan, chitin, mannan, β-(1, 5) galactofuranose, galactomannan (GM) and galactomannoprotein. These components are cross-linked with several other proteins. The cell wall is anchored with the underlying cell membrane by glycosylphosphatidylinositol (GPI)-binding protein. Sometimes outside the cell wall, a hydrophobic layer is found both in hyphae and conidia.

The conidia or asexual spores are pigmented, *echinulate*, *globose* to subglobose. In addition to the hydrophobic layer outside the cell wall, the conidia also contain a melanin layer. The sialic acid, composed of unsubstituted N-acetyl neuraminic acid linked with galactose by α-2, 6 bond, is detected in the conidial surface. The sialic acid helps in attachment of the conidia with the extracellular matrix. The conidia are produced in large numbers, and due to their small size (2–3 μm) and hydrophobicity, they can remain viable in the air for prolonged period. The mean concentration of *Aspergillus* conidia in air is 0.2–15 conidia/m^3 and is up to 10^6 conidia/m^3 in some agricultural settings. In the favourable environment, the conidia can swell and germinate into the hyphae.

Most of the *Aspergillus* does not have any sexual stage of the growth (Fungi imperfecti) except *A. fumigatus* (teleomorph *Neosartorya fumigata*), *A. oryzae* and *A. nidulans* which can produce sexual spores (ascospores). In both *A. fumigatus* and *A. oryzae*, two opposite mating types (idiomorphs, MAT1-1 and MAT1-2) are commonly found near close vicinity with the same frequency of occurrence (1:1). The MAT1-1 idiomorph is associated with increased virulence and higher resistance against antifungals.

4.4.2 Classification

Aspergillus belongs to Trichocomaceae family under the order Eurotiales and Phylum Ascomycota. The family contains another important genus known as *Penicillium*. Currently, the Trichocomaceae family is proposed to be further divided into three subfamilies such as Aspergillaceae, Trichocomaceae and Thermoascaceae. The genus *Aspergillus* contains more than 250 species grouped into subgenera and species complex. There are eight such subgenera, i.e. *Aspergillus*, *Fumigati*, *Circumdati*, *Candidi*, *Terrei*, *Nidulantes*, *Warcupi* and *Ornati*. The important species complexes are *A. fumigatus*, *A. flavus*, *A. terreus*, *A. niger*, *A. nidulans* and *A. ustus*. Molecular studies revealed several cryptic species (hidden species that are morphologically indistinguishable) within the *Aspergillus* species complex. For example, *Neosartorya peudofischeri*, *Aspergillus lentulus*, *A. viridinutans* and *A. fumigatiafinnis* are cryptic species described under *A. fumigatus* species complex.

4.4.3 Susceptibility to Disinfectants

Aspergillus is susceptible to disinfectants such as 35 % ethanol (*A. niger*) and copper-8-quinolinolate (0.4 mcg/mL). *A. niger* is highly sensitive to radiation such as gamma and X-ray (500 and 100 Krad), whereas lower dose of radiation (1–5 Krad) can stimulate the sporulation of the fungi.

4.4.4 Natural Habitat and Distribution

Aspergillus is commonly found in soil, air, water (fresh and marine), indoor environment (dust, above the false ceiling in a room) vegetables and feed as saprophytes. Some unusual habitats are also observed for *Aspergillus* (*A. carneus*, *A. flavus*, *A. nidulans*, *A. niger*, *A. terreus*) such as 'microbial mat' under the fresh water or hypersaline water (0–35 % salt concentration). They are common contaminants in the laboratory.

A. flavus follows a life cycle consisting of a saprophytic phase which occurs in soil, plant debris, tree leaves, decaying wood, animal fodder, cotton, compost piles, dead insect and animal carcasses, outdoor and indoor air environment (air ventilation system), stored grains and even in human or animal patients. The fungi can also act as opportunistic pathogen of plants and animals including human. In the saprophytic phase, they are present either as mycelium or a heavily melanised survival structures called *sclerotia*. In favourable environment, they produce conidiophores with conidia which are transmitted by air or any other means into the plants. After the establishment of the infection in the susceptible host, aflatoxins are produced. It is most prevalent in warm regions, especially between latitudes 35 N and 35 S. Other species of *Aspergillus* are worldwide in distribution.

4.4.5 Genome

The genome of *Aspergillus* (*A. flavus*) contains eight chromosomes. The genome size is 36.3 Mbp with 13,071 predicted genes. A 75 Kbp region in the chromosome consisting 29 gene clusters is responsible for the aflatoxin biosynthesis. Although there are some genes present outside the cluster which also helps in aflatoxin biosynthesis. The cluster is composed of a single gene encoding polyketide synthases (PKS), five genes encoding cytochrome P450 monooxygenases and genes for global regulation [mitogen-activated protein kinase (MAPK), MAPK kinase (MAPKK) and MAPKK kinase (MAPKKK), signal transduction, pathogenicity, oxidative stress and fungal development. The genes encoding for different enzymes such as amylase, cellulase, pectinases, proteases, chitinase, chitosanases and pectin methylesterases have been also identified in the genome.

The genome of *A. fumigatus* is 29.4 Mbp containing 9630 genes, most of them have unknown function. Deep sequencing study of *A. fumigatus* mRNA (RNA seq) revealed tens of unannotated and hundreds of novel genes encoding small proteins which are involved in colony growth and establishment of different clinical forms of aspergillosis.

4.4.6 Isolation, Growth and Colony Characteristics

Aspergillus can be isolated in Sabouraud dextrose agar (SDA) with or without chloramphenicol (0.05 g/L) and other common bacteriological media such as blood agar. The cycloheximide can inhibit the growth. The plates are incubated at 37 °C for 4–5 days. The species which cannot tolerate this temperature can be incubated at 25 °C. Some species, such as *A. fumigatus*, is thermophilic which is able to grow at 55–75 °C.

The colonies of *A. fumigatus* are white cottony which becomes velvety or granular with green colouration after several days (Fig. 4.7).

Whereas *A. flavus* colonies are primarily cottony and later become 'sugary texture' with yellowish green colour, *A. niger* colonies are white in colour in the primary stage which later becomes black due to production of black-coloured conidia. Similarly, *A. terreus* colonies are initially white which later become cinnamon-buff coloured with 'sugary texture'.

4.4.7 Antigenic Characteristics

Galactomannan (GM) is the major antigen of *Aspergillus* which is a carbohydrate molecule composed of mannose with side chains of β (1, 5)-linked galactofuranosyl residues. It is a part of cell wall along with chitin. It is released through the pores at the growing hyphal tips during logarithmic growth phase of the fungi in highest amount which helps in the detection of the antigen for diagnosis. GM is found in other fungi including *Penicillium*, *Fusarium*, *Alternaria*, *Histoplasma* and yeasts including *Cryptococcus* which can produce antigenic cross-reaction.

The cell wall of Aspergillus contains 1, 3 β D glucan (BDG) which is also secreted during late logarithmic growth phase of the fungi. The BDG is present in the cell walls of many pathogenic fungi including *Candida*, *Cryptococcus*, *Fusarium* and *Pneumocystis* producing antigenic cross-reaction.

Heat shock protein (Hspl/Asp f 12) is also detected as an immunodominant antigen in *A. fumigatus* which plays role in protective immunity.

4.4.8 Toxins

The mycotoxins are generally produced during hyphal growth and conidiogenesis which are secreted in the environment or released after the death of the hyphae. Conidia may also contain the preformed mycotoxin. *A. fumigatus* produces many toxins as secondary metabolites.

4.4.8.1 Aflatoxins

Aflatoxins are difuranocoumarin compounds produced by different species of *Aspergillus* as

Fig. 4.7 Growth of *Aspergillus fumigatus* in SDA

secondary metabolite. It has six major types such as B1, B2, G1, G2, M1 and M2. Aflatoxin M1 and M2 are produced as metabolites of B1 and B2, and they are commonly found in milk and to some extent in eggs. Among different species *Aspergillus flavus* is the chief producer which can affect many agricultural crops such as maize, cotton, groundnuts (peanuts), Brazil nuts, pecans, pistachio nuts and walnuts. Both preharvest and postharvest contamination of these crops with aflatoxins is common. However, other species such as *A. fumigatus* can produce aflatoxin B1 and G1.

Among the six types of aflatoxin, B1 (AFB1) is the most potent toxin and carcinogen than others in human and animals including birds, fish and rodents. The chronic intoxication can cause immunosuppression, malnutrition, centrilobular necrosis and fatty infiltration of the liver including hepatomas and degeneration of respiratory system. Maximum permissible limit of total aflatoxin and M1 (AFM1) is 20 ppb in feed and food for human and 0.5 ppb in milk, respectively, as guidelined by the US Food and Drug Administration (FDA). European Commission has set the limit at 15 ppb for total aflatoxin in groundnuts.

The aflatoxin synthesis genes are located by making a cluster in the genome of *A. flavus*. The major regulator gene is *aflR*, located in the centre of the cluster. Major transcription regulator gene is *aflS* (*aflJ*) located adjacent to the *aflR*. Other unrelated genes such as *laeA* and *veA* are involved in global regulatory role in aflatoxin biosynthesis. Many nutritional factors such as carbon, nitrogen, amino acid, lipid and trace elements control aflatoxin biosynthesis. Other regulatory factors required for aflatoxin synthesis includes the temperature (28–35 °C), acidic pH, hot weather and drought conditions, sporulation and sclerotia formation of the fungi and oxidative stress. Certain plant metabolite such as n-decyl aldehyde is inhibitory to aflatoxin production.

4.4.8.2 Gliotoxin

It is the major and most potent toxin produced by 95 % strains of *A. fumigatus* isolated from both clinical cases and environment. It belongs to the family of epipolythiodioxopiperazines, characterised by a disulphide bridge across a piperazine ring which is essential for their toxicity. It can inhibit macrophage and neutrophil-mediated phagocytosis, mitogen-activated T cell proliferation, mast cell activation, cytotoxic T cell response and reactive oxygen species production. Further, the gliotoxin causes monocyte apoptosis, inhibition of epithelial cells ciliary movement and epithelial cell damage. There is a putative 12-gene cluster which encodes the toxin. Among the gene cluster, *gliZ* and *gliP* play major role.

4.4.8.3 Ochratoxins

In tropical countries, ochratoxins are mainly produced by *A. ochraceus* and *A. carbonarius*. *A. ochraceus* are mostly associated with dried and stored foods (cereals). Whereas *A. carbonarius* is commonly found in fruits (grapes) that mature in sunlight and at high temperature, the ochratoxin is considered as potent nephrotoxic, carcinogenic (human carcinogen of 2B group), genotoxic and immunotoxic.

4.4.8.4 Citrinin

It is a fungal metabolite which was first isolated from *Penicillium citrinum*. Other fungi such as *A. terreus* can also produce citrinin. It is frequently detected in feed stuffs or food along with ochratoxin A, and it can increase the toxicity of ochratoxin synergistically. The intoxication with both of the toxins together may cause endemic nephropathy. Further, citrinin is also embryocidal and foetotoxic.

4.4.8.5 Ribotoxins (Restrictocin, Mitogillin)

The ribotoxins are produced by *A. fumigatus* which specifically targets the sarcin/ricin domain of 28S rRNA and inhibits the protein synthesis. The toxins are encoded by *asp fl/mitF/res* gene. Mitogillin is highly cytotoxic causing cell death even in low concentration.

4.4.8.6 Haemolysin

A. fumigatus produces a haemolysin which is encoded by *aspHS* gene. The toxin can lyse the rabbit and sheep red blood cells (RBC), macrophages and endothelial cells.

4.4.8.7 Helvolic Acid and Fumagillin

Helvolic acid belongs to a group of steroidal antibiotic, known as fusidanes. It is produced by *A. fumigatus*. It can inhibit the oxidative burst of macrophages and metabolism of low-density lipoproteins, and it can induce ciliostasis and rupture of epithelial cells.

Fumagillin is an antitumour antibiotic which can inhibit angiogenesis, endothelial cell proliferation and ciliary movement in the respiratory epithelium.

4.4.8.8 Fumitremorgin (A,B,C) and Tryptoquivaline A

These are neurotropic toxins produced by some strains of *A. fumigatus*. They can cause tremor, seizure and abnormal behaviours in experimental mice.

4.4.9 Virulence Factors

The virulence factors possessed by *Aspergillus* are described in Table 4.11.

4.4.10 Transmission

Inhalation of contaminated dust or ingestion of contaminated feed with the fungal spores is the major route of transmission in animals and poultry. Intramammary inoculation of spores may cause mastitis in cattle. In poultry farm, the spores are introduced with the contaminated feed or litter. In moist environment, they germinate and produce more numbers of spores which are inhaled by the birds. The poor ventilation and sanitation, humidity and long-term storage of feed in a farm may increase the spore concentration in the air. The predisposing factors are impaired immunity or stressed condition of the birds due to administration of antibiotic or steroids, vaccination, metabolic bone disease, inadequate diet resulting hypovitaminosis A, overcrowding, shipping, starvation, thermal discomfort, inbreeding, toxicosis, traumatic injuries and reproductive activity.

4.4.11 Pathogenesis

4.4.11.1 *A. fumigatus*

Healthy animals and human are exposed to the *Aspergillus* spores because the mean concentration of conidia in air is 0.2 to $15/m^3$ and is up to $10^6/m^3$ in some agricultural settings. In spite of this constant exposure, the animals or human do not suffer regularly which suggests that inhalation of spores in large numbers is required to establish the infection. After entry within the host through inhalation, the conidia encounter with the respiratory mucosa comprised of mucus, proteins, lipids, ions, water and other cellular secretions that contribute to the mucociliary clearance. The conidia are engulfed by tracheal epithelial cells, alveolar type II cells and nasal epithelial cells. Most of them are killed by the phago-lysosome, although the remaining conidia are capable to germinate and exit the phago-lysosome. Conidial sialic acids act as ligand for adherence with the alveolar epithelial cells especially with fibrinogen and fibronectin, commonly found in the wounded epithelial surfaces. So lung injury acts as major predisposing factor for causation of invasive aspergillosis. Further, the gliotoxin, fumagillin and helvolic acid produced by the fungi cause damage in the respiratory mucosa and slow ciliary movement facilitating the attachment of the conidia.

After penetration of the respiratory mucosa, the conidia are deposited into the alveolar space. In the poultry, air sac is the primary target organ of the fungi because the inhaled air reaches the air sacs prior to the epithelial surfaces of the lungs. Later the conidia reach the lung parenchyma and are embedded in the atria or infundibulum. The conidial maturation begins which causes loss of hydrophobic layer and exposure of the inner cell wall. The cell wall is composed of galactomannan, chitin and β-glucan which act as ligand for the soluble and cell-associated pattern recognition receptors (PRR). The soluble receptors are a family of C-type lectins including lung surfactant proteins A and D and mannan-binding lectin (MBL). They act as opsonin and can bind the *A. fumigatus* cell wall carbohydrates

Table 4.11 Virulence factors of *Aspergillus*

Virulence factors	Function
Conidia	Due to its small size, thermotolerance, resistance to oxidative stress and high growth rate, they can easily transmit and establish the infection within the host. The pigmentation of the conidia can reduce C3 complement deposition and neutrophil activation
Autophagy related gene (ATG)	Autophagy helps the fungi to survive under the nutrient stress condition by helping in conidiation, hyphal foraging and maintenance of metal-ion homeostasis
α-1,2-mannosyltransferase (encoded by *kre2 afmnt1*)	A putative virulence factor related with thermotolerance (48 °C) of *A. fumigatus*. In mutants, cell wall instability is observed which causes leakage at the hyphal tips and inhibition of growth at 48 °C
cgrA (nucleolar protein, encoded by *cgrA*)	It helps in ribosome biogenesis at 37 °C. The mutants are less virulent
Heat shock protein (Hsp1/Asp f 12)	It acts as molecular chaperone and helps in protein transport and growth at 37 °C. It can produce stress response during inflammation
Fks1 (encoded by *fks1*)	It is the catalytic subunit of glucan synthase enzyme involved in synthesis of β-glucan, essential for the fungal growth
Chitin synthase (encoded by *chsA*, *chsB*, *chsC*, *chsD*, *chsE*, *chsF*, *chsG*)	chsG is the only virulence factor involved in synthesis of chitin. The mutants show reduced virulence
Galactomannan (GM)	It helps in adhesion to host components such as fibronectin, laminin, pentraxin3 and other surface receptors of macrophages, dendritic cells and langerhans cells. The released GM is used in commercial antigen detection kit. The galactomannan along with the α-glucans forms a polysaccharide matrix which helps in formation of biofilms. The biofilms resist the antifungals and concentrate the enzymes produced during the growth which helps in further tissue invasion
Glycosylphosphatidylinositol (GPI) anchor	The GPI anchor is required for cell wall integrity, morphogenesis and virulence of the fungi
Phosphomannose isomerase enzyme (Pmi1)	It is required for both cell wall synthesis and energy production in *A. fumigatus*, but its actual role in virulence is not elucidated
Rodlet	Rodlet is a microfibril filled with proteins (hydrophobins), encoded by *rodA*, *rodB* and others. These proteins help in conidial dispersion, adhesion of the hyphae with the soil surface or lung epithelium
Melanin	It is a pigment which protects the genome of conidia from ultraviolet light, enzymatic lysis and oxidation. It also helps in survival of the conidia in the environment. In vivo, the melanin can protect the conidia from complement-mediated lysis and reactive oxygen species
Catalases (catA, Cat1/catB, catC, catE and cat2) and superoxide dismutases (SODs cytoplasmic Cu/Zn SOD (Sod1),mitochondrial MnSOD (Sod2),cytoplasmic MnSOD(Sod3) and (Sod4)	It helps in detoxification of reactive oxygen species (ROS) produced by macrophages and neutrophils
Protease (alkaline protease, elastase), aspergillopepsin (aspartic protease, Pep), Dipeptidyl peptidases (Dpp)	They degrade the structural barrier (collagen, elastin, fibrinogen, casein) in the host to obtain the nutrients The Dipeptidyl peptidases (Dpp) can degrade collagen, hormones and cytokines. They have T cell modulation property

(continued)

Table 4.11 (continued)

Virulence factors	Function
Phospholipase (A–D)	They can break the ester bond of phosphoglycerides and thus destabilised the cell membrane to cause cell lysis
Trehalose biosynthetic proteins (encoded by *tpsA*, *tpsB*)	They are associated with germination of conidia at 37 °C and inhibition of oxidative stress
Siderophores (fusaricine C, triacetylfusaricine C, ferricrocin and hydroxyferricrocin, encoded by different *sid* genes)	They are low molecular weight proteins (Mr < 1,500), involved in iron uptake for the fungi. They are ferric iron specific and high affinity chelators. The siderophores are also required for germ tube formation, asexual sporulation, resistance to oxidative stress, catalase activity and virulence
Zinc transporter protein (encoded by *ZrfA-C*)	It affects the germination of conidia and growth capacity under zinc-limiting conditions
cAMP-PKA (cAMP-protein kinase A) signalling pathway (encoded by *pkaC1* gene)	The mutants are hypovirulent having reduced growth, germination rate and increased susceptibility to oxidative stress
Ras proteins (RasA, RasB, RhbA)	Ras proteins are monomeric GTPases which act as molecular switches that transduce signals from the outside of the cell to signalling cascades inside the cell. The *rasB* mutants are hypovirulent having reduced growth and germination rate
Histidine kinase (fos1, tcsB)	It is involved in signal transduction under stress condition. The fos1 mutants are remarkably hypovirulent
Calcium signalling pathway (calA/cnaA)	It is involved in stress response, mating, budding and tolerance to antifungal drugs. The cnaA mutant has decreased filamentation, defective conidia and reduced virulence
Cross pathway control (CPC, *cpcA*) signalling pathway	This signalling pathway is involved in amino acid homeostasis under environmental stress. The *cpcA* mutants are hypovirulent
Transcription factor Ace2	It is associated with conidia formation, pigmentation and virulence
Development-regulated protein (MedA)	It is associated with adherence, host interaction and virulence
Unfolded protein response (UPR)	It is a stress response sensor associated with protein-folding activities in endoplasmic reticulum. Among the three sensors (IRE1, ATF6 and PERK) identified in eukaryotes, IREA is so far reported from *A. fumigatus*, *A. niger* and *A. nidulans*. It is regulated by *ireA*, *hacA* genes. The mutants show increased cell wall fragility, impaired secretory system that is unable to extract nutrients from host tissues and lack of growth in hypoxic condition (*ireA* mutant)
Dol-P-Man:protein *O*-mannosyltransferases (PMT)	This enzyme helps in *O*-mannosylation of the proteins in endoplasmic reticulum which is required for the construction and maintenance of the cell wall. Loss of *pmt* gene is lethal for *A. fumigatus*
Ubiqutin encoding genes (*UBI1*, *UBI4*)	They help in stress and metabolic adaptation

in a calcium-dependent manner. This binding enhances phagocytosis by the alveolar macrophages, whereas the cell-associated receptors such as toll-like receptors (TLR2, TLR4) and dectin-1(ligand β-glucan) are involved in recognition of conidia and hyphae. Most of the conidia are killed by reactive oxygen species (ROS) or acidified phago-lysosome produced within the alveolar macrophages. Non-oxidative mechanism may also contribute in killing of the conidia. In immunosuppressed animals having defective alveolar macrophage function, the conidia are able to escape the phago-lysosome-mediated killing. The virulence factors of *Aspergillus* such as melanin, rodlets and superoxide dismutase can protect the conidia from reactive oxygen species. Other blood cells such as platelets can also damage the *A. fumigatus* conidia by releasing serotonin.

The conidia which survive from this first line of defence can germinate. The germination involves conidial swelling (isotropic growth) followed by protrusion of germ tube (polar growth). The polar growth occurs along with mitotic nuclear division followed by a septum formation at the base of the germ tube. Several such septi are formed after completion of each mitotic cycle to create elongating hyphae which can invade the tissues. This germination depends on nutrient sensing, acquisition and biosynthetic pathways for obtaining the nutrients from the host. Under the nutrient-limited condition and other stress factors, the protein-folding capacity of the fungal endoplasmic reticulum reaches its maximum limit and the unfolded protein response (UPR) initiates. The enzymes such as elastase, aspartic proteinase, serine protease and metalloprotease produced by *Aspergillus* help in the breakdown of host tissue for nutrient acquisition and invasion. For survival and virulence of the fungi in vivo, there is a requirement for biosynthesis of uracil/uridine, folate and lysine. Aspergilli can also utilise a wide range of nitrogen sources in which environmental nitrate is taken and converted into ammonium and subsequently glutamine and glutamate. They degrade the host proteins to obtain amino acids and other nutrients or they can synthesise the amino acids from carbon- and nitrogen-containing precursors. This synthesis is regulated by cross pathway control (*cpcA*) locus. *A. fumigatus* can also uptake the iron with the help of siderophores or through the reductive iron assimilation. Besides nutritional requirements, *A. fumigatus* also adapts with the in vivo environment by the expression of different factors such as pH regulator (PacC) for adaptation in alkaline pH, sterol-regulatory element-binding protein (SrbA) for adaptation in hypoxic condition, etc. Production of different toxins such as gliotoxin, ribotoxins and haemolysin also help in the establishment of infection by preventing cellular effector function and causing cellular apoptosis.

A. fumigatus growing hyphae invades the endothelial cell lining of blood vessels (angioinvasion) from the albuminal side to the luminal side. The angioinvasion commonly occurs during neutropenia. During the invasion, part of hyphae may break off and invade the endothelium of the opposite side. Haematogenous dissemination of the hyphae into vital organs may take place. During angioinvasion, thrombosis and infarction occur creating a necrotic focus (plaque) where the fungi can grow. The hyphae are not effectively phagocytosed by the macrophages due to their larger size. The neutrophils aggregate around the hyphae and damage them by reactive oxygen species and a variety of antimicrobial compounds such as proteases, defensins, lysozyme and lactoferrin released from their granules. The neutrophils can also inhibit the hyphal growth by the lactoferrin sequestration of iron, an essential nutrient for fungal growth. The dying neutrophils produce an extracellular trap with the nuclear DNA and several antifungal proteins which can also inhibit the hyphal growth.

4.4.12 Disease Produced

The major animal and human diseases produced by different species of *Aspergillus* are enlisted in Table 4.12.

4.4.13 Immunity

The innate immune system recognises *Aspergillus* through either soluble receptors [C-type lectins, mannose-binding lectin (MBL), pentraxin 3 (PTX3)] or the cell-associated pattern recognition receptors (TLR 2, 4 and dectin1). Recently, intracellular NOD-like receptor (NOD2 and NLRP3) is also found to recognise *Aspergillus*. Generally the cell wall carbohydrates of the fungi

Table 4.12 Major diseases of animals and human caused by *Aspergillus* sp.

Fungi	Host	Disease
A. fumigatus, *A. flavus*, *A. niger*, *A. glaucus*, *A. nidulans* (*A. fumigatus* is the major cause probably due to smaller size of the spores than other species which helps in easy transmission)	Poultry (poultry are relatively more susceptible due to differences in anatomical, physiological and respiratory immune system compared with other mammals)	*Avian aspergillosis/brooder pneumonia* The disease is characterised by rhinitis, mycotic keratitis associated with blepharospasm, photophobia, periorbital swelling, turbid discharge, swollen and adhered eyelids, cloudy cornea and cheesy yellow exudates within the conjunctival sac, lung congestion, unilateral drooping of the wing due to infection of the thoracic air sac, clavicular air sac or the proximal humerus, repeated vomition due to lesions in the anterior air sacs. The respiratory signs are not regularly detected. In postmortem examination, green/yellow/white necrotic foci are observed in the lung and other organs
A. fumigatus	Horse	(a) *Guttural pouch mycosis* Necrotizing inflammation and formation of diphtheritic membrane in guttural pouch (auditory tube diverticulum). Clinical signs include epistaxis, unilateral purulent nasal discharge, abnormal head position. It occurs most commonly in stabled horses during summer (b) Nasal granuloma (c) Corneal ulcer
A. fumigatus	Dogs	(a) *Canine sinonasal aspergillosis/canine rhinitis* The disease is characterised by persistent sneezing and nasal discharge which is unresponsive to antibiotics. The infection involves paranasal sinuses, nasal cavity, turbinate bones and CNS. Golden Retrievers and Collie breeds are more susceptible (b) Otitis externa (c) Chronic rhinitis
A. fumigatus	Human	*Invasive aspergillosis* A disease of the respiratory tract which involves the lung parenchyma, pleura, trachea and bronchi. It is common in the patients with haematological malignancy, prolonged antibiotic users or stem cell transplant recipients
A. fumigatus	Cattle, horse	Sporadic abortion especially during winter which is associated with hay feeding. The fungi invade the foetus and produce cutaneous lesions
A. fumigatus	Cattle	(a) Mastitis (abscess in udder) (b) Pneumonia
Aspergillus sp.	Calves	Mycotic gastritis
Aspergillus sp.	Horses	*Keratomycosis* It is characterised by corneal opacity, ulceration and formation of endothelial plaques in the centre of the cornea

(β-glucan, GM, chitin) act as ligand (pathogen-associated molecular pattern, PAMP). The soluble receptors act as opsonin which enhances phagocytosis by the alveolar macrophages. Pentraxin (PTX3) specifically binds the conidia (not hyphae) and enhances their phagocytosis by dendritic cells and alveolar macrophages. After the compatible binding of the fungi and cell-associated PRRs, the cytoplasmic tail attached with the receptors which contains an immunoreceptor tyrosine-based activation (ITAM)-like motif gets phosphorylated. This phosphorylation activates several pathways to stimulate the phagocytes for killing of the conidia through the production of NADPH-dependent reactive oxygen species (ROS). The phosphorylation also activates the NFκB and stimulates the secretion of pro-inflammatory cytokines (IL12, TNFα). The cytokines help in the recruitment of the immune cells at the site of the infection. Further, *A. fumigates* hyphae (not conidia), through the production of ROS, can stimulate NLRP3 inflammasome to release active IL1β which can recruit more amount of neutrophils. These neutrophils produce a neutrophil entrapment traps (NETs) to engulf the larger hyphae which are not phagocytosed by the macrophages.

The innate immune system also influences the subsequent development of T cell-mediated adaptive response. The adaptive response is initiated with the dendritic cells which can engulf the conidia or hyphae to carry them into the draining lymph nodes. The conidial engulfment occurs through the 'coiling phagocytosis' in which the dendritic pseudopods rotate around the conidia to form a layer following the binding of the mannose receptor DC-SIGN and complement receptor 3 (CR3). Whereas the hyphal engulfment occurs through 'zipper phagocytosis' after the binding of Fc receptor (as ligand) and complement receptor 3 (CR3), this kind of phagocytosis occurs following the contour of the fungi which further activates the IL4-producing CD4 T cells in the spleen and mediastinal lymph nodes. The dendritic cells can also recognise the damage-associated molecular pattern (DAMP) of the fungi through the receptor for advanced glycation end products (RAGE) to activate the T cells. The epithelial cells in addition express the pattern recognition receptors for fungal recognition and as a consequence induce the T cell tolerance.

The activated Th1 cells secrete IFN γ which helps in the clearance of the hyphae and produces a protective immunity. Whereas activation of Th2 cells causes the secretion of IL4, IL5 is responsible for fungal allergy in the host and it prevents the clearance. Th17 cell activation correlates with *Aspergillus* infection during immunodeficiencies. The Th17 cells secrete IL17 and IL22 which recruits neutrophils. Further, fungal growth, inflammatory immunity and tolerance to *Aspergillus* are controlled by the activation of natural (n) Treg cells. This kind of Treg cells can suppress the neutrophils through the combined actions of IL10 and CTLA4 acting on dendritic cells to produce indoleamine 2, 3-dioxygenase (IDO). This enzyme helps in controlling inflammation, infection and allergy. Thus, protective immunity to *Aspergillus* is dependent on balanced act of Treg cells over effector Th1 and Th17 cells, with the contribution of IDO.

4.4.14 Diagnosis

4.4.14.1 Clinical Specimens

The clinical specimens from animals or poultry include mastitic milk, foetal stomach contents, ear swabs, skin scrapping and biopsies from nasal granuloma and plaques in the guttural pouch. The specimen from human is lower respiratory tract sample by bronchoalveolar lavage. After postmortem, pneumonic lung, nodules from vital organs may be collected. The small tissue samples may be wrapped in a piece of paper before putting into the 10 % formalin.

4.4.14.2 Laboratory Examination

1. *Direct examination*:

   The clinical specimens especially the tissue scrapings should be cleared with 10 % KOH and are observed under the microscope.

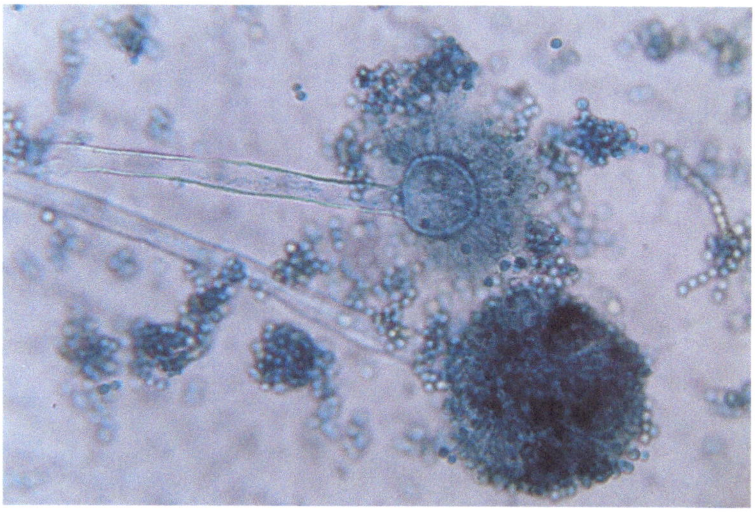

Fig. 4.8 Vesicle of *Aspergillus flavus* (Lactophenol cotton blue, 100×)

Histopathological staining can be performed with periodic acid–Schiff (PAS), Grocott's silver or methenamine silver stain for detection of tissue invasion. In the tissue section, *Aspergillus* hyphae are narrow and septate which are not easily distinguishable from other fungi.

Different species of *Aspergillus* has characteristic fruiting body structures through which they can be identified by an experienced person. *A. fumigatus* conidiophores have a foot cell at their bases. The dome-shaped vesicles contain phialides in their upper portion. From the phialides, long chain of conidia is borne which may sweep inwards, whereas in case of *A. flavus*, the vesicles are round in shape with the phialides containing spores present on the entire surface (Fig. 4.8).

In *A. niger*, vesicles are spherical which bear large metulae supporting small phialides. The conidia are black in colour.

2. *Isolation and identification*: Media and incubation condition as described earlier will serve the purpose. However, it is the most common fungal contaminant in the laboratory and it is routinely isolated from respiratory tract of healthy animals. So repeated isolation from the clinical specimen is required, along with correlation with the history, clinical signs and histopathological observations for proper diagnosis of clinical aspergillosis.
3. *Detection of antigen*:
   Galactomannan (GM) is the predominant antigen released by *A. fumigatus* in the circulation during angioinvasion which can be detected by latex agglutination test, sandwich ELISA in serum or urine samples. The detection limit of both the tests is 15 ng/mL and 1 ng/mL, respectively. However, the GM detection assay is not specific for *Aspergillus* as it is cross-reacting with other fungi such as *Penicillium, Fusarium, Alternaria* and *Histoplasma*. Further, recent investigations revealed the presence of GM in tazobactam antibiotic preparations, nutritional supplements and electrolyte solutions which may produce false-positive reaction especially in human.

   Similarly, 1, 3 β D glucan (BDG) can be detected for identification of *Aspergillus*. However, it is produced by a lot of other fungi such as *Candida, Fusarium, Pneumocystis*, etc. So the test can predict the general fungal infection rather than specifically aspergillosis. There are several BDG assay kits available for use in human such as Fungitell kit using amoebocyte lysates from *Limulus polyphemus* and Fungitec-G kit using reagents from *Tachypleus tridentatus*.
4. *Molecular biology*: Confirmation of *A. fumigatus* isolates by PCR-RFLP can be performed using *BccI*, *MspI* and *Sau3AI* restriction enzymes.

4.4.15 Treatment

In poultry the treatment can be performed by topical application of antifungal after removal of thoracic granuloma followed by systemic antifungal therapy. However, removal of granuloma from the respiratory tract is difficult. In that case, only topical application can be performed by nebulisation, nasal or air sac flushing, or surgical irrigation of the abdominal cavities. For topical application, amphotericin B (1 mg/kg body weight, 10–14 days, nebulisation), enilconazole (0.1 mL/kg body weight, 5 days, nebulisation), miconazole and terbinafine can be used, whereas for systemic application, amphotericin B (1.5 mg/kg body weight, 3–5 days, intravenous), fluconazole (5 mg/kg body weight, 7 days, oral or intravenous), itraconazole (5–15 mg/kg body weight, 5–21 days, oral), ketoconazole (10–30 mg/kg body weight, 21 days, oral), voriconazole or 5-fluorocytosine can be used.

Resistance to antifungals is increasingly reported especially from human patients since 1997 when the first published case of itraconazole-resistant *A. fumigatus* appeared. The itraconazole and other triazole compounds bind lanosterol (encoded by *CYP51*) and prevent the synthesis of ergosterol from it. Point mutation in *CYP51* is detected to be responsible for the resistance. However, resistance to amphotericin B is limited among the *Aspergillus* isolates except *A. niger*.

4.5 *Blastomyces*

In 1894, Thomas C. Gilchrist first recognised blastomycosis in a patient with cutaneous lesion in Baltimore, USA. In his honour, the infection was named as Gilchrist's disease. He first described the causative agent as a protozoon which was disproved by him and Stokes when they isolated the fungus from another cutaneous lesion in 1898 and named it *Blastomyces dermatitidis*.

In India, *Blastomyces dermatitidis* was first reported from tea garden workers having wart-like superficial lesions at the border area of Cooch Behar (West Bengal) and Bhutan (Ganguly 1925). Panja (1925) also diagnosed a case of generalised blastomycosis in scrapings from the cutaneous nodular lesions. Further, the blastomycosis is reported from insectivorous bats (Khan et al. 1982), male Mongrel dog (Iyer 1983) and migratory birds (Rawal et al. 1988).

4.5.1 Morphology

Blastomyces dermatitidis is a thermally dimorphic fungus exhibiting the mycelial form at room temperature and yeast form at 37 °C. The mycelial form consists of the branching hyphae which are 2–3 μm in diameter. The conidiophores carrying single terminal conidia are present at the right angle of the hyphal branches which resembles lollipops (Fig. 4.9). The conidia are round or oval in shape, 2–10 μm in diameter and prefer to enter the susceptible host through inhalation.

The yeast cells are usually 8–15 μm in diameter, encapsulated with a thick refractile cell wall and contain 8–12 nuclei. They reproduce by a single broad-based bud, with the daughter cell often as large as the mother cell before detachment (Fig. 4.10). The bud is attached with the mother cell with a persistent cell wall and a wide pore between them. This characteristic feature is useful for identification of the yeast in the clinical specimens.

The mycelium to yeast phase conversion depends on increased temperature and hybrid histidine kinase (DRK1) which acts as

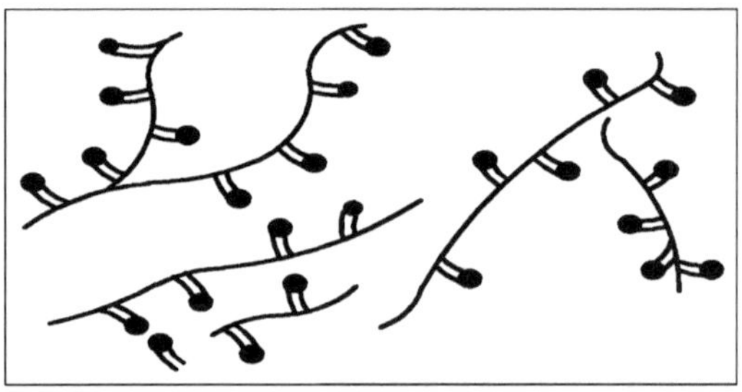

Fig. 4.9 Conidial lollipop-like arrangement of *Blastomyces dermatitidis* (schematic diagram)

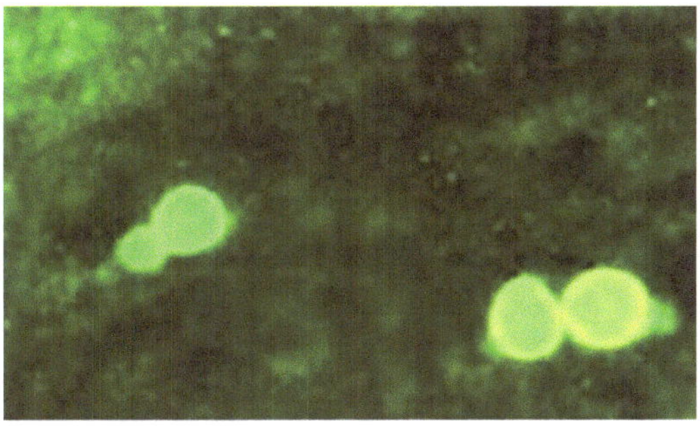

Fig. 4.10 Budding characteristics of *Blastomyces dermatitidis* in a FAT smear (Photograph courtesy: Prof. P.P. Gupta, Ex-Director of Veterinary Research, Punjab Agricultural University, Punjab, India)

two-component signalling system. The conversion results in an increased cell wall content of α (1, 3)-glucan and a decreased β (1, 3)-glucan which is associated with increased virulence. Multiple changes occur in the lipid composition of plasma membrane which causes remodelling and reorganisation of the membrane. The yeast-specific protein such as BAD1 (formerly WI-1) is expressed.

4.5.2 Classification

The genus *Blastomyces* belongs to the phylum Ascomycota, class Euascomycetes, order Onygenales and family Onygenaceae. The sole pathogenic species under the genus is *Blastomyces dermatitidis*.

4.5.3 Reproduction

B. dermatitidis can reproduce sexually and the sexual state (teleomorph) is known as *Ajellomyces dermatitidis*. It is generally considered as heterothallic species, and both the mating types (plus and minus) are required to complete the sexual cycle. The mating types occur in equal frequency in the nature. However, the presence of homothallic mating system is also detected. The mating results a thick-walled spore-bearing structure (*cleistothecia*) and surrounding numerous asci. Each ascus contains eight ascospores. The mating begins with the formation of dikaryon which is followed by the production of ascogenous hyphae. These hyphae develop into asci in which meiosis, recombination and mitosis yield eight haploid ascospores.

It is not clear whether sexual reproduction of *B. dermatitidis* occurs in the nature and whether the generated spores help in transmission of the infection.

4.5.4 Susceptibility to Disinfectants

B. dermatitidis are susceptible to 1 % sodium hypochlorite, phenolics, glutaraldehyde, formaldehyde, 10 % formalin.

4.5.5 Natural Habitat and Distribution

Blastomyces prefer to inhabit in warm, mild acidic, moist, sandy soils with high organic content such as animal droppings and in close proximity to water bodies. The presence of moisture helps in release of asexual conidia which can enter the susceptible host to establish an infection. Outbreaks of blastomycosis have also been associated with disruptions of the soil, such as excavation that might lead to the increase number of spores or hyphal fragments in the air. The fungi are also isolated from the poultry house, mule stall and pigeon droppings, indicating their relationship with animal excreta.

Indian fruit bat (*Pteropus giganteus*) is a natural reservoir of *B. dermatitidis*. However, regarding other bat species as a reservoir of *Blastomyces* is currently unknown.

The natural reservoir of the fungus is geographically restricted to a specific area (microfocus). So the outbreaks occur in that area or due to travel or fomite-mediated transmission from that area. In the United States, most clinical cases of blastomycosis occur surrounding the Mississippi and Ohio rivers such as Arkansas, Kentucky, Mississippi, North Carolina, Tennessee, Louisiana, Illinois and Wisconsin in addition to regions of Canada and United States bordering the Great Lakes and St. Lawrence

River. It is also endemic throughout Africa especially in 16 countries from Algeria to South Africa. The sporadic infections are observed in Israel, Saudi Arabia and India.

4.5.6 Genome

The sexual mating is governed by a single *MAT* locus with two idiomorphs. Each *MAT* idiomorph encodes either an alpha domain or high mobility group (HMG) domain transcription factor. The alpha domain and HMG domain genes determine plus and minus strains, and the corresponding *MAT* locus is designated as *MAT1-1* and *MAT1-2*, respectively. These genes that determine the mating types are evolutionarily conserved. The *MAT* locus is flanked by the *SLA2*, *COX13* and *APN2* genes.

4.5.7 Isolation, Growth and Colony Characteristics

Both yeast and mycelial forms of *B. dermatitidis* can be isolated in the laboratory with suitable media and incubation temperature. The preferred culture media for the mycelial form are Sabouraud dextrose agar, potato dextrose agar and potato flake agar. Media such as the brain–heart infusion agar with blood and chloramphenicol, yeast extract phosphate agar and inhibitory mould agar are selective media which can inhibit the growth of saprophytic fungi and contaminating bacteria. Some strains of *B. dermatitidis* are able to grow in cycloheximide-containing medium. So both the media with and without cycloheximide can be used.

At 25 °C, *B. dermatitidis* can take up to 4–6 weeks to form typical mould colonies that appear white to off-white and glabrous or waxy and becoming grey to brown as aerial hyphae develop with age. The aerial hyphae as spicules may be visible after 1 week incubation.

At 37 °C, the hyphae convert slowly into yeast cells. The colonies are typically white to beige, creamy and 0.5–3 cm in diameter.

4.5.8 Antigenic Characteristics

The yeast form of *B. dermatitidis* possesses the antigen-A (135 KDa), composed of protein and carbohydrate (37 %). On the basis of its presence, *B. dermatitidis* has two major serotypes such as A-antigen positive and A-antigen negative. Among them the A-antigen-negative serotypes are restricted within Africa.

Klein and Jones (1990) identified another immunodominant antigen of *B. dermatitidis* yeasts (WI-1) which contributes in humoural and cellular immunity. It is a 120 KDa protein which is present in the yeast cell surface but not the hyphal filaments or the conidia. It also acts as adhesin and it contains the 24 amino acid tandem repeat which is 90 % homologous with the *Yersinia* adhesin. The carboxy terminal of WI-1 contains a cysteine-rich domain that is similar to epidermal growth factor and it mediates binding to extracellular matrix. Due to this binding property, the protein is renamed as *Blastomyces* adhesin 1 (BAD1).

The carbohydrate moiety of *B. dermatitidis* A-antigen is shared with other fungi such as *Histoplasma* which creates cross-reactivity. However, due to very low level of carbohydrate (less than 1.5 %), the WI-1 is not cross-reactive with others. So it is a better choice as a diagnostic antigen.

4.5.9 Virulence Factors

The virulence factors possessed by *B. dermatitidis* are described in Table 4.13.

4.5.10 Transmission

Inhalation of spores or mycelial fragments is the major route of transmission of *B. dermatitidis* in animals and human. The zoonotic transmission is rare and the person-to-person transmission does not occur. However, cutaneous or percutaneous infection occurs in human through the infected dog bite. The veterinarians may get

Table 4.13 Virulence mechanisms and factors possessed by *B. dermatitidis*

Virulence factors	Functions
α (1, 3)-glucan	(i) It conceals immunostimulatory β-glucan layer of the cell wall from detection by host phagocytic cells (ii) It inhibits the production of the pro-inflammatory cytokine (TNFα) by phagocytes
BAD1(WI-1) adhesin	(i) It helps in attachment of the yeast with macrophages which is mediated through complement receptor (CR3) (ii) The BAD1 suppress the production of TNFα by neutrophils and macrophages via a mechanism that involves upregulation and secretion of TGFβ which can further downregulate the Th1 response. It also impairs the complement cascades in the peripheral blood. It helps in the progression of pulmonary infection (iii) The BAD1 contains 35 copies of a 25 amino acid tandem repeat which can bind calcium and helps the fungi to grow in calcium poor condition
Phospholipid	The strains of *B. dermatitidis* lacking phopsholipids in the cell wall are less virulent than the wild types. The presence of phospholipid produces granulomatous tissue reaction which is not lethal and necrotizing
Alkali-soluble cell wall fraction	Lethal, tissue necrotising, endotoxin-like activity
DOPA melanin (eumelanin)	It is synthesised in fungi by polyketide synthase and/or phenoloxidases pathway which requires the presence of exogenous substrate such as O-diphenolic and P-diphenolic compounds. It is produced by conidia and yeast phase, not the hyphae. It is observed that melanin-producing yeast cells are less susceptible to antifungals

the infection during the postmortem examination of infected dog without proper precaution. Unusually human cases of blastomycosis are reported, originated from sexual contact and intrauterine transmission.

A number of occupational and recreational activities such as forestry, construction, gardening, fishing, hunting and hiking have been associated with a risk of infection. These kinds of activities increase the exposure to the natural reservoir of *B. dermatitidis*. Seasonal variations of blastomycosis have been observed. The infection is more common in the late summer, autumn and winter months due to ideal temperature for mycelial growth, increased outdoor activities and heat stress. All age groups are susceptible to infection; however, the mean age in human affected with blastomycosis is detected as 20–70 which is consistent with the outdoor activities of this age group.

Sexually active male dogs aged between 2 and 4 years, larger breeds (Hounds) and immunocompromised dogs are more susceptible. Residing or travelling to the endemic zones act as additional risk factors.

4.5.11 Pathogenesis

After transmission through the inhalation route, the conidia are deposited in the lungs where natural resistance is mediated by neutrophils, monocytes and alveolar macrophages that can phagocytose and kill the conidia. The macrophages can also inhibit the conidia to yeast conversion. If the inhaled conidia overcome the natural resistance and are converted into the yeast form, they are difficult to phagocytose due to thick capsule. The neutrophils are also relatively inefficient in phagocytosing the yeasts. The proliferation of the yeasts occurs asexually by budding. The proliferating yeasts may produce the infection with the help of the virulence factors as described in Table 4.13.

The infection may further progress through the lymphohaematogenous route. In decreasing order of frequency, the skin, bones and male genitourinary system are the most common sites where the extrapulmonary infection may disseminate. With the development of immunity inflammatory pyrogranulomatous, reaction occurs either in the primary site of infection or in any

Table 4.14 Major diseases of animal and human caused by *Blastomyces dermatitidis*

Fungi	Host	Disease
Blastomyces dermatitidis	Dogs	Lung lesions occur in 65–85 % of cases and may be clinically silent or associated with cough, tachypnea, exercise intolerance, cyanosis or respiratory distress. Other major clinical signs are anorexia, focal bone infections causing lameness and osteolysis, lymphadenopathy, skin lesions, corneal opacity, conjunctivitis, blindness, orchitis and immunosuppression. Neurologic symptoms are uncommon in dogs reported in only 3–6 % of cases. Reported neurologic signs include lethargy, neck pain, circling, cranial nerve deficits, head pressing, seizures, hypermetria, ataxia and tetraparesis
	Cats	Pyrogranulomatous inflammation of different organs, nasal discharge, cough, weight loss
	Horses	Pyrogranulomatous pneumonia, pleuritis, peritonitis, cutaneous abscess
	Human	The clinical signs are highly variable. Common signs include weight loss, fever, fatigue and other nonspecific complaints. Pulmonary infection produces either acute or chronic pneumonia. The acute pneumonic patients show fever, chills and productive purulent cough, with or without haemoptysis. The patients with chronic pneumonia show weight loss, night sweats, fever and cough with sputum production. The extrapulmonary infection is characterised by skin lesions which are either verrucous (raised lesion with sharp and irregular border) or ulcerative, osteomyelitis, prostatitis, orchitis and meningitis

other organ-producing granulomas. These granulomas do not caseate as occurs in case of tuberculosis. After recovery from the pulmonary infection, sometimes endogenous reactivation occurs with or without previous therapy.

4.5.12 Disease Produced

The major animal and human diseases produced by *Blastomyces* are enlisted in Table 4.14.

4.5.13 Immunity

The cell-mediated immunity plays major role in effective protection against *B. dermatitidis*. It is primarily mediated by T lymphocytes (Th-1 response) and lymphokine-activated macrophages. Experimentally successful transfer of immunity occurs in mice against *B. dermatitidis* through the T cells.

However, the antibodies against *B. dermatitidis* can also effectively kill the fungi through the activation of complements. The BAD1 adhesin of *B. dermatitidis* is an immunodominant antigen that produces both the cellular and antibody-mediated responses. Experimentally the monoclonal antibody to BAD1 significantly enhances the C3-binding capacity of *B. dermatitidis*, although the passive transfer of antibodies could not protect the experimental animals infected with *B. dermatitidis*.

4.5.14 Diagnosis

4.5.14.1 Clinical Specimen

The clinical specimens for diagnosis of blastomycosis includes bronchoalveolar lavage, saliva or sputum, skin scrapings, exudates from the lesion, synovial fluid, urine, cerebrospinal fluid, bone marrow, fluid from anterior chamber of the eye and lymph node aspirates. For cutaneous lesion, the samples should be collected from the active part, and the biopsy specimen should include the full thickness of the lesion.

4.5.14.2 Laboratory Examination

1. *Direct examination*: The smear prepared from the clinical specimens can be observed under microscope by 10 % KOH preparation for demonstration of yeast. The KOH dissolves the tissue and cell debris present in the clinical specimens especially in the saliva and skin scrapings to clear the visibility. The yeast cells with the broad-based bud are the diagnostic character. The KOH can be used along with calcofluor white, a fluorochrome compound that binds to chitin in the walls of fungal cells and make them visible under fluorescence microscope. Currently the yeast cells

are also demonstrated by the negative staining with the Diff-Quik (DQ) stain which demonstrates the unstained yeast cells present in the stained background. Since they are not commensal and colonisation does not occur so their detection in direct smear establishes the diagnosis, and it is considered as most rapid diagnostic technique for blastomycosis. However, negative finding is inconclusive as 37–46 % of human blastomycosis patients reveal the presence of *Blastomyces* in direct smear.

2. *Isolation and identification*: Media and incubation condition as described earlier will serve the purpose. It is the most sensitive and gold standard method for diagnosis of blastomycosis. However, the mycelial form is highly infectious, and the caution should be taken in handling of the culture in the laboratory and it is a time-consuming process causing delay in diagnosis. The confirmation of culture can be done by conversion of yeast phase into the mould phase, PCR and commercially available DNA probe.
3. *Histopathology*: In the clinical specimens such as saliva or sputum, fine needle aspirates and formalin-fixed tissues, the classical look of the *Blastomyces* yeasts such as round to oval multinucleate cells with a single broad-based bud located intra or extracellularly helps in diagnosis. The tissue sections are stained with haematoxylin and eosin, periodic acid–Schiff stain (PAS), PAS with haematoxylin counter stain and Grocott methenamine silver (GMS). If the classical look is not detected, then the diagnosis becomes difficult as it produces confusion with other yeasts. In such cases, the mucicarmine stain can differentiate the *Cryptococcus neoformans* and *Blastomyces* as it can specifically stain the capsule present surrounding the *Cryptococcus* cells. Melanin stain can also differentiate *Blastomyces* from other yeasts. The Alcian blue or acid-fast stain can be used for differentiation between *Blastomyces* and *Coccidioides immitis*. *Blastomyces* is weakly positive, whereas *Coccidioides immitis* is negative. In situ hybridisation with specific probe can diagnose *Blastomyces* within the tissues.
4. *Detection of antigen*: In human, the ELISA-based kits are available to detect the *Blastomyces* antigen (A-antigen) in urine, cerebrospinal fluid and bronchoalveolar lavage fluid. Sometimes it produces cross-reaction with *Histoplasma capsulatum* var. *capsulatum* due to the presence of shared carbohydrate fraction of A-antigen. Currently BAD1 is the antigenic target having no cross-reaction with *Histoplasma* as it does not contain carbohydrate. Detection of β-D-glucan (BDG) is not so effective in blastomycosis.
5. *Serological tests*:
   (a) Complement fixation test (CFT) can be performed with the yeast phase antigen for detection of anti-*Blastomyces* antibodies. However, the sensitivity and specificity is low (57 and 30 %, respectively).
   (b) Immunodiffusion using purified *B. dermatitidis* A-antigen is more specific and sensitive (65–80 %).
   (c) Enzyme-linked immunossorbent assay (ELISA) and radio immune assay (RIA)-based techniques are currently most sensitive techniques for the detection of *Blastomyces* antibodies. The sandwich and competitive-binding inhibition ELISA are used for detection of antibodies.
6. *Skin test (delayed type of hypersensitivity)*: The crude fungal extract (blastomycin) can be inoculated into the susceptible animal for detection of hypersensitivity (like tuberculin test). However, the test is neither sensitive nor specific. The swelling rapidly diminishes and the test is negative during repeat testing.
7. *Molecular biology*: PCR-based technique such as nested PCR can be used for detection of *Blastomyces*.

4.5.15 Treatment

In dogs and human, treatment with amphotericin B lipid complex and itraconazole is safe and effective. The amphotericin B is found effective in respiratory distress, CNS involvement and in children with blastomycosis. However, amphotericin B deoxycholate has relatively high rate of toxicity such as infusion reaction, nephrotoxicity, hypokalaemia, hypomagnesaemia and anaemia. Amphotericin B lipid complex is safer but is clinically as effective as the deoxycholate.

Itraconazole is the drug of choice for non-meningeal, non-life-threatening blastomycosis in immunocompetent patients. However, it has severe drug–drug interactions especially with antiretroviral and antirejection drugs. Fluconazole drug–drug interaction is less, and it can easily penetrate the CNS with better bioavailability in oral route, but it has limited efficacy against *Blastomyces*. The second-generation triazole compounds such as voriconazole and posaconazole are promising against blastomycosis.

Currently a primary treatment with amphotericin B lipid complex followed by any azole compound preferably voriconazole is recommended to treat blastomycosis.

4.6 *Coccidioides*

In 1892, Alejandro Posadas, a medical student and pathologist Robert Wernicke, first described *Coccidioides* from a soldier with recurrent skin tumours in Argentina. They described it as unknown parasite found in the lesion having similarity with *Coccidia*. In 1986, Rixford and Gilchrist also identified it as a coccidial agent and designated it as *Coccidioides immitis* (immitis = not mild). Its fungal identity was confirmed in 1900 by Ophüls and Moffitt. They further showed the organism was dimorphic which switched from the mycelial phase in culture to spherule forming phase in the tissues. In 1937, Ernest Dickson established the clinical importance of *Coccidioides immitis* when it was identified as the etiological agent of 'Valley fever' at San Joaquin Valley of California. He also described a milder form of infection which occurred in his laboratory during accidental opening of the Petri dish with the mould by a medical student. Dickson coined the term 'coccidioidomycosis' to include all forms of infection by *Coccidioides*.

4.6.1 Morphology

Coccidioides is a dimorphic fungus having two phases in the life cycle. The saprophytic phase occurs in the environment and artificial laboratory culture, and the parasitic phase occurs in the host such as human, dogs, horses and other animals.

The saprophytic phase consists of mycelia with septate hyphae which are 2–4 μm in diameter. In 5–7 days, the hyphae yield a chain of multinucleate arthroconidia along the length of the hyphae (enteroarthric development). Initially, small, nonviable, brittle cells are produced, known as disjunctors. These cells degenerate to release the buoyant *arthroconidia* which act as the source of infection especially during soil disruption and storm.

Within the body of the host, the arthroconidia are converted into a different morphologic form, known as *spherule*, in the presence of CO_2 and phagocytic cells. The arthroconidia shed the outer wall and enlarge to produce the immature spherule containing one nucleus. The central portion is occupied with a large vacuole. The nucleus undergoes continuous division which is followed by partitioning of the cytoplasm surrounding the vacuole. It produces matured spherule (60–100 μm diameter) containing 800–1,000 endospores (2–5 μm in diameter each). The spherule wall contains glucans, chitin, mannose polymers, 3-*O*-methylmannose and galactose, whereas the wall of the endospores is composed of glucan and chitin. The spherules release the endospores in vivo which again enlarge to form a new spherule filled with endospores to repeat the cycle. The mechanism of release is unexplored, although probable role

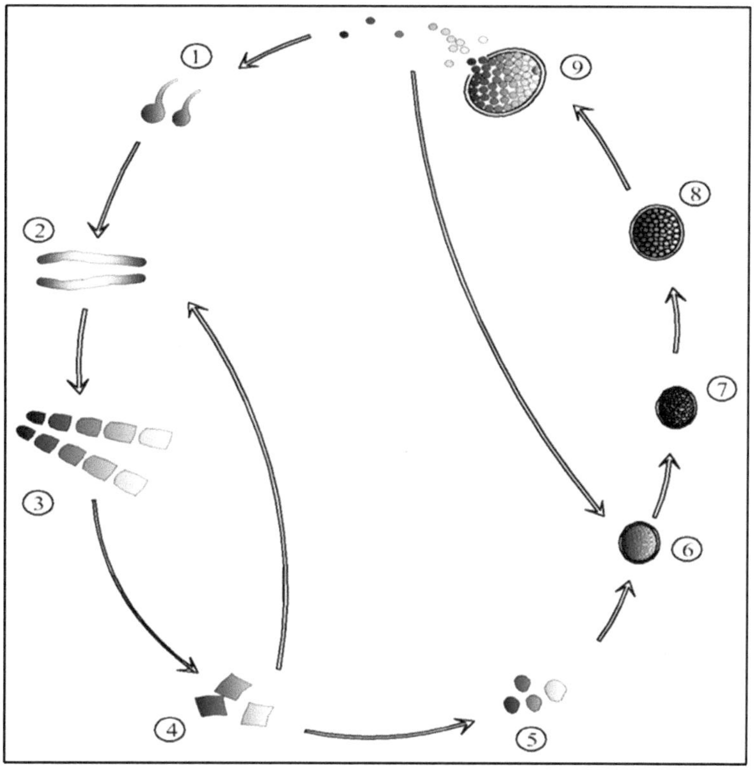

Fig. 4.11 Life cycle of *Coccidioides immitis* (schematic representation). *1* Emergence of tubular structure, *2* formation of mycelia, *3* fragmentation of hyphae, *4* release of arthroconidia, *5* enlargement of arthroconidia within the host body, *6*, *7* spherule formation, *8* endospore generation within spherule, *9* release of endospores from spherule

of certain enzymes such as chitinase and glucanase is observed. The endospores may further generate the mycelial form if they return to the soil or grow in artificial culture medium (Fig. 4.11).

4.6.2 Classification

The genus *Coccidioides* belongs to the phylum Ascomycota, class Euascomycetes, order Onygenales and family Onygenaceae like *Blastomyces* and *Histoplasma*. Previously *Coccidioides immitis* was considered as the sole pathogenic species under the genus. The phylogenetic analysis using single nucleotide polymorphisms and microsatellites has showed the existence of two genetically different *C. immitis* clades, California and non-California. Later the California clade was designated as *C. immitis* and non-California clade was designated as a new species, namely, *C. posadasii*, in honour of Alejandro Posadas. There are minor differences in amino acid sequence between two species, but they are similar in antigenicity, virulence or morphology.

4.6.3 Reproduction

No sexual stage (teleomorph) of the fungi is detected.

4.6.4 Susceptibility to Disinfectants

C. immitis is susceptible to 70 % ethanol for 1–20 min exposure, 37 % commercial formaldehyde, chlorine dioxide gas (sporicidal), 1:10 dilution of bleach, 6 % hydrogen peroxide and 3 % phenolics with a contact time of 20 min or more. For fumigation of laboratory with paraformaldehyde, 0.3 g/ft^{3}of room air is recommended to make it *C. immitis* free. The mycelia form can be inactivated by heat at 120 °C for 30 min.

4.6.5 Natural Habitat and Distribution

Moist alkaline soil rich in salt and carbonised organic materials is required for growth of the hyphal phase. Following a subsequent dry period, the hyphae will die leaving the arthrospores. Disturbance of the soil by different human or animal activities such as construction, agriculture, excavation and archaeological exploration is required for transmission of the spores into the body of the host.

There are a few endemic zones where *C. immitis* are naturally found. The high endemic zones include Southwestern United States and the bordering regions of northern Mexico. In the United States, south central portion of Arizona, particularly Tucson and Phoenix, the southern one-third of California, notably the San Joaquin Valley, and southwestern Texas are most affected. The climate in this region is arid with a yearly rainfall ranging from 10 to 50 cm, with extremely hot summers, winters with few freezes and alkaline, sandy soil.

The sporadic human cases are reported from Central and South America, Utah, Nevada, Brazilian states (Piauı', Bahia, Ceara', Maranha˜o), Guatemala, Honduras, Nicaragua, Argentina, Paraguay, Colombia and Venezuela.

In India, a few human cases of Coccidioidomycosis are reported which are imported in nature. The patients either travelled or lived for a certain period in the vicinity of Arizona, United States. However, in animals, no authentic report of Coccidioidomycosis is available till date.

4.6.6 Genome

The genome of *C. immitis* (28.9 Mbp) and *C. posadasii* (27 Mbp) contain 10,355 and 7,229 number of annotated genes, respectively. The mean GC% of genome is 46. The mean intron and exon number per gene is 2.3 and 3.4, respectively. *Coccidioides* species are estimated to have four chromosomes by CHEF gel analysis. Like other Ascomycetes, the chromosome structure of *Coccidioides* also contains species-specific 'genomic islands', located in chromosomal sub-telomeric regions.

The phylogenetic analysis with the ancestor and close relative of *Coccidioides* revealed the gain of several virulence genes when they are diverged from their close relative. These acquired genes help the fungi to grow within the host even in the carcasses. One of such acquired gene encodes spherule-associated enzyme related with allantoin metabolism which acts as nitrogen source in the host and source of proteins for haeme-/tetrapyrrole binding required for scavenging iron within a host. However, the analysis also detected reduced numbers of gene encoding plant degrading enzymes (cellulases, cutinases, tannases, pectinesterases) and proteins associated with sugar metabolism in the genome of *Coccidioides* in comparison to other Ascomycetes. It indicated the association of the fungi with live animals rather than the plants or soil.

Comparative genomic analysis between the two species of *Coccidioides* revealed that the 8 % of the genes in the *C. immitis* population may be introgressed from *C. posadasii*. This introgression is driven by the natural selection because the introgressed region is enriched with coding sequences.

4.6.7 Isolation, Growth and Colony Characteristics

Coccidioides can be isolated in ordinary fungal media (Sabouraud dextrose agar, inhibitory mould agar, brain–heart infusion agar, potato dextrose agar, potato flakes agar) even in the presence of cycloheximide as well as in bacterial media (blood agar, chocolate agar, selective yeast extract medium used for *Legionella* or *Bordetella*) when incubated at 25 °C or 30–37 °C for 4–5 days (range is 2–16 days). *C. posadasii* was reported to grow faster than *C. immitis* at 37 °C. The selective medium along with cycloheximide is used to isolate *Coccidioides* from mixed flora.

The spherule is formed at 40 °C in the media containing casein hydrolysate, glucose, biotin, glutathione and salt.

The young colonies are white and moist, glabrous and tenacious, and adhering to the medium. At this young stage, no arthroconidia is detected. With maturity of the culture, discrete concentric rings and the filamentous area containing arthroconidia become visible. The pigmentation with tan, brown, pink, grey or yellow colour was also detected in matured colonies.

4.6.8 Antigenic Characteristics

The major immunodominant antigen of *C. immitis* is spherule outer wall glycoprotein (SOW Gp) which can produce both antibody-mediated and cellular immune responses. This glycoprotein consists of a signal peptide and pro-peptide, a proline and aspartic acid-rich tandem repeat motif, and a glycosylphosphatidylinositol (GPI) anchor. The repeat domain is 41–47 amino acids in length and contains three to six copies of proline and aspartic acid-rich sequences. The number of copies varies with strain. However, the antigen is shared by genetically and geographically diverse strains.

The saprobic phase contains a heat-stable, 19 kDa antigen with proteinase activity which is also considered as a *Coccidioides*-specific antigen (CS-Ag). Recently, an in-house antigen of *Coccidioides* is described containing β-glucosidase (45–67 KDa) and glutamine synthetase (67–97 KDa) as major immunoreactive proteins which can react with serum samples of the patients suffering from the infection.

The carbohydrate antigen is detected in the mycelial extract (coccidioidin) along with some amino acid nitrogen.

4.6.9 Virulence Factors

The virulence factors possessed by *Coccidioides immitis* are enlisted in Table 4.15.

Table 4.15 Virulence mechanisms and factors possessed by *C. immitis*

Virulence factors	Functions
Extracellular proteinases	The soluble conidial wall fraction (SCWF) is antiphagocytic, immunosuppressive for mice lymph node proliferation. The fraction contains chiefly a serine proteinase enzyme, a dimer of 60KDa. It is released from the outer cell wall of the conidia after inhalation, to damage the respiratory tract since elastin is the major structural component of the tissues. Two other proteinase enzymes (56 Kda and 19 Kda) are also described. The 56 KDa proteinase is unable to degrade immunoglobulin, whereas the 19 KDa fraction (*Coccidioides*-specific antigen, CS-Ag) is able to degrade immunoglobulin
	The proteinases help in the release of endospores from the spherule and local tissue damage. The elastin degradation products attract inflammatory cells to produce damaging inflammatory response
Oestrogen-binding proteins	Certain pregnancy-related hormones such as progesterone, testosterone, 17β-estradiol (E2) stimulate the rate of spherule maturation and endospore release of *Coccidioides*. The oestrogen-binding proteins present in the cytosol of the fungi help in this process. In general, men are more susceptible than women for the disseminated infection. However, during pregnancy the risk of disseminated infection increases in women especially during late pregnancy which is directly correlated with increased E2 level
Urease (encoded by *ure*)	(a) It maintains the alkalinity of the host tissue to produce favourable condition for the fungal growth (b) It also makes the phagosome (containing endospores) alkaline to prevent its fusion with lysosome (c) It increases the expression of arginase I. The arginase I competes with the inducible nitric oxide synthase (iNOS) produced by the host phagocytes for the same substrate (L-arginine). Thus, arginase can inhibit the production of nitric oxide which is lethal to the fungi
Spherule	The spherule is antiphagocytic. Due to its larger size, it escapes the phagocytosis by the neutrophils and macrophages
Spherule outer wall glycoprotein	It acts as adhesin in colonising respiratory tract

4.6.10 Transmission

The infection is transmitted through inhalation of arthrospores which are easily deposited into the lung due to their smaller size. A single spore is sufficient to produce an infection in human. The chances of getting more than one spore are associated with the disturbance of soil rich in mycelial form especially during military manoeuvres or archaeologic excavation. Rarely, cutaneous lesions can develop as the result of direct inoculation.

Immunosuppressed individuals such as patients of acquired immunodeficiency syndrome (AIDS), Hodgkin's disease, malignant neoplasms, uraemia, collagen-vascular diseases or those who had received recent immunosuppressive drug therapy and organ or bone marrow transplant recipients are more susceptible. Pregnancy especially during the third trimester and the peripartum period also increases the risk of getting the infection. Rarely the neonatal coccidioidomycosis has been reported which is maternal in origin. Certain human races such as Filipinos, African Americans, Mexican Americans and Native Americans are more susceptible to the infection associated with genetic factors. There is no known human-to-human or animal-to-human transmission of this infection.

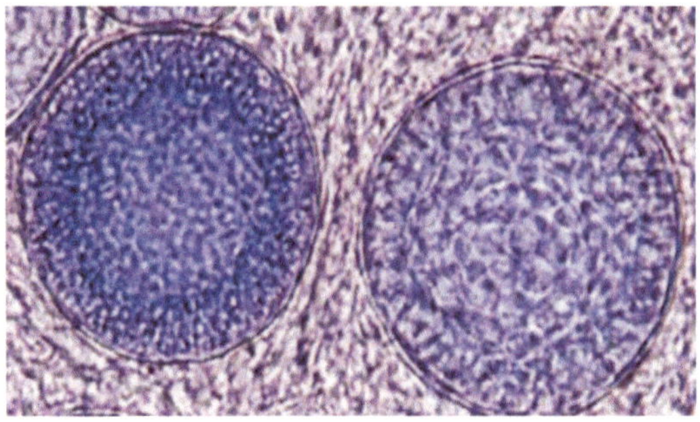

Fig. 4.12 Spherule of *Coccidioides immitis* in stained tissue smear (Photograph courtesy: Prof. P.P. Gupta, Ex-Director of Veterinary Research, Punjab Agricultural University, Punjab, India)

4.6.11 Pathogenesis

After transmission within the body of the host, the arthroconidia are converted into spherule with numerous endospores as depicted earlier (Fig. 4.12). The spherule can escape the phagocytosis by the host neutrophils, macrophages and dendritic cells due to their larger size (60–100 μm). The protease present in the spherule can digest the antibody or other opsonins to evade phagocytosis. Further, the spherules showed moderate resistance against reactive oxygen species (ROS) such as hydrogen peroxide produced by the host macrophages. The spherules also inhibit the production of nitric oxide (NO) by the host phagocytes. So, even after phagocytosis, they can inhibit the oxygen-mediated killing. However, the spherule outer wall glycoprotein (SOWGp) can bind antibody to enhance opsonisation and subsequent phagocytosis. The peak of SOWGp production occurs during isotropic growth of the spherule but then suddenly decreases during endospore formation. During the endospore forming phase, a 34 kDa metalloprotease (Mep1) that belongs to the metzincin superfamily is produced which can digest the SOWGp. So, the Mep1 plays a crucial role in the establishment of the infection.

The endospores released from the spherule can be phagocytosed, but many of them survive due to antiphagocytic nature of the endospore outer wall. Some of them, even after phagocytosis, can survive by employing various strategies. They can secrete the urease enzyme which elevates the arginase I level within the phagosome. The arginase I can increase polyamine synthesis and can decrease the nitric oxide (NO) production which is favourable for the fungal survival. The alkalinisation of the microenvironment as a consequence of urease activity prevents the fusion of phagosome with lysosomes.

Production of anti-*Coccidioides* Th1 immunity (cell-mediated immunity) offers resistance, whereas Th2 immunity (antibody mediated) is associated with the establishment of infection. Experimentally the secreted endosporulation antigens (EA) of *Coccidioides* in high concentration can suppress the Th1 immunity, whereas in

Table 4.16 Major diseases of animal and human caused by *C. immitis*

Fungi	Host	Disease
Coccidioides immitis	Dogs	In dogs, primary pulmonary or disseminated coccidioidomycosis is developed. Signs of pulmonary form include coughing, weight loss, fever, lethargy and anorexia. The pneumonia is either mild interstitial or extensive miliary pneumonia with consolidation of the lung lobes or only hilar lymphadenopathy (enlargement of lymph nodes within the lung hilum). The hilar lymphadenopathy is considered as a significant mark for diagnosis. The bone involvement is restricted within appendicular skeleton and is characterised by bone lysis as well as periosteal new bone formation. The CNS involvement is characterised by seizure
	Cats	Like dogs, the cats also suffer from both the pulmonary and disseminated form. The CNS form is characterised by incoordination, hyperesthesia, seizures and behaviour changes. The skin lesions include dermatitis, ulceration or abscesses that do not respond to antibiotics
	Horse	Disseminated coccidioidomycosis is observed. The female horses suffer more than male. The infection is characterised by abortion, mastitis, osteomyelitis, nasal granuloma and meningitis
	Llama	Pulmonary and disseminated Coccidioidomycosis is observed which is characterised by respiratory distress, dermatitis, osteomyelitis, meningitis, polyperiarthritis syndrome
	Cattle, sheep	Self-limiting infection is observed which is confined within the bronchial and mediastinal lymph nodes
	Human	*California disease/desert fever/Posadas-Wernicke disease/San Joaquin fever/valley fever*: Primary infections affect respiratory system. Influenza-like, pneumonic and pleural presentations are the most common. The disseminated form is noticed causing tissue damage in the skin, bone, joints and central nervous system

low concentration stimulates the same. This finding shows that both inhibitory and stimulatory components are present in EA and the establishment or clearing of the infection is dependent on their ratio.

After establishment of the infection in absence or reduced Th1 immunity, the endospores may disseminate via haematogenous (fungaemia) or lymphatic drainage into the bone, skin, lymph nodes, central nervous system (CNS) and heart. In dogs, the bone is the most common site of dissemination especially in the appendicular skeleton. In human, vertebral osteomyelitis is most prevalent. The CNS manifestation in animals is largely due to the production of granulomatous mass in the brain, not due to meningitis which is typical in human. Pericardial dissemination is much common in dogs than human.

The disseminated infection may remain dormant for many years in human (25–35 % cases). The infection recurs rarely due to immunosuppression, malnutrition, advancement of age and high dose of exposure. In animals, the recurrence is not reported but cannot be ruled out.

4.6.12 Disease Produced

The major animal and human diseases produced by *Coccidioides* are enlisted in Table 4.16.

4.6.13 Immunity

Primary recognition of the fungi is mediated through pattern recognition receptors (PRRs) expressed by the host immune cells. Among them, toll-like receptor (TLR2), dectin1, mannose receptor (MR) and surfactant proteins A (SPA) and D (SPD) are involved in recognition of *Coccidioides* ligands (pathogen-associated molecular pattern, PAMP) (Table 4.17).

The professional phagocytes especially the polymorphonuclear leukocytes (PMN/neutrophils) and macrophages are the first line of defence, recruited at the site of Coccidioidal infection. They try to clear the pathogen through oxidative or non-oxidative killing mechanism. However, it depends on fungal strain and morphotype. The arthroconidia and endospores are more susceptible than spherule. The dendritic cells have the ability

Table 4.17 Pattern recognition receptors and their ligands involved in recognition of *Coccidioides*

Pattern recognition receptors (PRRs)	Coccidioidal pathogen-associated molecular pattern (PAMPs)
TLR2	Unidentified
Dectin1	β-glucan
Mannose receptor (MR)	Mannan
Surfactant proteins A (SPA)	Unidentified
Surfactant proteins D (SPD)	Unidentified

to bind and internalise the spherules. It is followed by production of a T cell-mediated response because they can elevate the T cell co-stimulatory molecule production.

The PAMP–PRR interactions stimulate the production of IL12 and IL23 which can convert the naive T cells into antigen-specific T cell populations. There are two functionally distinct subsets of CD4+ T cells, i.e. Th1 and Th2 cells. Both clinical and experimental evidence have demonstrated that Th1-mediated immunity is essential for defence against coccidioidomycosis. The activation of Th1 cells will increase the level of IFNγ, IL2 and other cytokines which provides the signals for recruiting and activating immune effector cells, whereas the activation of Th2 response increases the antibody level such as serum IgG, IgE and IgA which does not offer any protection.

4.6.14 Diagnosis

4.6.14.1 Clinical Specimens

The clinical specimens for diagnosis of coccidioidomycosis include pus or exudates from lesion, transtracheal washes, skin scrapings and biopsies from lymph node, bone, etc. In human, the specimens include sputum, bronchoalveolar lavage, transtracheal aspirates and lung biopsies. All specimens should be labelled as potentially hazardous during transport.

4.6.14.2 Laboratory Examination

1. *Direct examination*: The wet mounts with 10 % KOH are employed on the clinical specimen for detection of the spherule with endospores which is diagnostic of coccidioidomycosis. The presence of immature spherule or endospores without spherule may be confused with other fungal infection. The calcofluor white (CFW) fluorescent stain offers best detection of the fungi by binding the chitin and cellulose of the cell wall. However, it can also nonspecifically stain the plant and fatty substances.
2. *Isolation and identification*: Media and incubation condition as described earlier will serve the purpose. However, *Coccidioides* is extremely hazardous to handle, and several cases of infection from the laboratories are documented through inhalation of arthroconidia. It is also considered as a 'select agent' of bioterrorism. So, *Coccidioides*-specific laboratory should have biosafety level 3 containment maintained with negative air pressure. Further, expertise is needed in handling of the colonies because the arthroconidia are easily generated. However, the routine laboratory where suspected culture of *Coccidioides* is handled should have minimum biosafety level 2 facilities. All the cultures are handled under class II biological safety cabinet containing vertical laminar airflow with HEPA (high efficiency particulate air) filter and exhaust air. The lid of the Petri dishes along with suspected culture should be taped, if preserved for 5–7 days. After identification, the routine laboratory should destroy the culture properly. The culture should not be transferred without prior permission from the authority.
3. *Histopathology*: The histopathological examination can be performed on the biopsy materials collected from the skin, bone, lung, brain, etc., to confirm whether the infection is disseminated into different organs. In the tissues, endosporulating spherules, empty ruptured spherules and immature non-endosporulating spherules can be detected by staining. Sometimes, mycelia are observed in the tissues which are indistinguishable from other common fungi. The Grocott methenamine silver stain is the best stain for detection of *Coccidioides* in histopathological sections.

However, Grocott methenamine silver can also stain the tissue elements such as mucus, glycogen granules and some bacteria. It causes overstaining of the fungi which can mask the endospores within the spherule. Other histological stains, such as periodic acid–Schiff (PAS), haematoxylin-eosin, Giemsa, papanicolaou and mucicarmine, can also be used as a substitute.

4. *Detection of antigen*: Detection of antigen is useful in the very early stage of coccidioidomycosis when antibody is not developed (within 1 week) and in the patients suffering from hypogammaglobulinaemia or other immunodeficiency who are unable to produce the antibody. The tests such as radioimmunoassay (RIA) and inhibition ELISA were developed to detect the protein–carbohydrate conjugate antigen of the fungi.
5. *Serological tests*: The serological tests are not only diagnostic but also indicative about the prognosis of the infection. Detection of anti-*Coccidioides* IgM (within 1–3 weeks of infection) can be done by tube precipitin (TP) and immunodiffusion (ID) tests. The TP is based on heat-stable 120 kDa β glucosidase antigen, whereas detection of IgG (after third week–several months) can be done by complement fixation test (CFT) which is based on heat-labile chitinase antigen. However, canine serum contains anticomplementary substances; so CFT cannot be performed in the dogs. Serological IgG titre can indicate about the severity of the infection in human. For example, increasing IgG titre such as 1:32 or more indicates about the disseminated infection in human, whereas IgG titre between 1:2 and 1:16 is considered as positive for infected dogs.

   Sometimes, false-positive band is produced in ID test due to reaction between coccidioidal antigen and C-reactive substances present in the serum. This false band can be dissolved with 1.5 % aqueous solution of sodium EDTA, whereas the authentic band will remain as such even after washing.

   The cross-reaction is observed with other systemic fungi. The sera from mild and primary cases of coccidioidomycosis produce higher IgG titres with *Histoplasma* and *Blastomyces* antigens than with *Coccidioides* antigen.

   Currently ELISA-based tests are also developed for detection of anti-*Coccidioides* antibodies.
6. *Skin test for detection of delayed-type hypersensitivity (DTH)*: The antigen used in classical skin test is known as 'coccidioidin'. Primarily it was prepared as a soluble broth culture filtrate of mycelial cells grown for 2 months in a synthetic asparagine–glycerol–salt medium. Later, an aqueous lysate of spherules was prepared from *Coccidioides* strain (Silveira) that had been grown in Converse medium and then incubated in distilled water for up to 40 days at 34 °C. The soluble aqueous lysate is designated as 'spherulin' which can also be used as antigen in skin test. The skin test can be performed in human, dogs, horses, llamas and nonhuman primates. Reversion of positive result into the negative indicates grave prognosis. However, negative result cannot rule out the infection.
7. *Molecular biology*: Primarily a polymerase chain reaction (PCR) was developed using the primers of 18S rDNA of *Coccidioides*. Further, a real-time PCR is developed with primers amplifying a 170 base-pair part in the ITS2 (internal-transcribed spacer 2) region of the rDNA genome. Several studies also successfully validated about using PCR to detect the unique coccidioidal gene [antigen-2/proline rich antigen (Ag2/PRA)].

4.6.15 Treatment

In dogs and cats, oral treatment with ketoconazole, itraconazole and fluconazole with a dose range between 2.5 mg/kg twice daily to 20 mg/kg twice daily is usually performed to treat coccidioidomycosis. The adverse side effects of oral azole therapy in dogs and cats are similar with human. It includes reduced appetite, vomition and diarrhoea. Sometimes, a vasculitis causing suppurative skin lesions has been observed as an adverse effect of itraconazole in dogs. Similarly, polyuria/polydipsia, thinning of hair coat is detected with fluconazole treatment in dogs. Reduced fertility of male dogs is reported with ketoconazole treatment.

As an intravenous infusion, amphotericin B deoxycholate can be given. However, it has potent renal toxicity. Instead, currently amphotericin B lipid complex is preferred to treat the canine and feline patients who are not responding with the orally administered azoles.

Administration of oral antifungals is not so effective in llamas and alpacas where the drugs are metabolised and lost without absorption in the fermentative forestomach. Further, their jugular vein is also not easily accessible for intravenous application of antifungals.

4.7 *Histoplasma*

Histoplasma is an intracellular dimorphic fungus causing severe pulmonary and systemic infection in man and animals. *Histoplasma capsulatum* var. *farciminosum*, causative agent of Epizootic lymphangitis in horses, was first demonstrated in pus in 1873 by Rivolta. However, it was successfully isolated in 1896 by Tokishiga in Japan. The infection histoplasmosis in human was first notified by Darling, an American pathologist who observed it during autopsy of a Martinique native person died with tuberculosis-like syndrome in Panama (Darling 1906). He also described two similar cases observed in a Cantonese and another Martinique native, both of whom had resided in Panama for a long period. Initially the fungal aetiology was not ascertained; rather it was thought as 'encapsulated protozoa residing within histiocytes', so it was named as *Histoplasma*. In 1934, William DeMonbreun established the association of histoplasmosis with fungus. He first isolated the organism in artificial media, showed its dimorphism and experimentally produced the infection in laboratory animals to meet the conditions of Koch's postulate (De Monbreun 1934). In 1939, he described the infection in a dog died from hepatomegaly (De Monbreun 1939).

4.7.1 Morphology

Histoplasma is a classical example of thermally dimorphic fungus, which exhibits two different morphotypes (i.e. yeast and mould) in two kinds of temperature regime. The pathogenic yeast form is observed within the host tissue (37 °C), whereas the invasive mould form prevails in the environment (25–30 °C).

In the infected tissues, the oval yeast cells of *H. capsulatum* var. *capsulatum* are observed within macrophages, 3–5 μm in diameter. Their cell wall does not allow the easy passage of Gram or Giemsa stain, so the cell is observed to be surrounded with an empty areola in a stained smear, producing the impression of a capsule (Fig. 4.13). They are rarely found in extracellular condition and they infrequently show the budding. They can be stained by Gomori

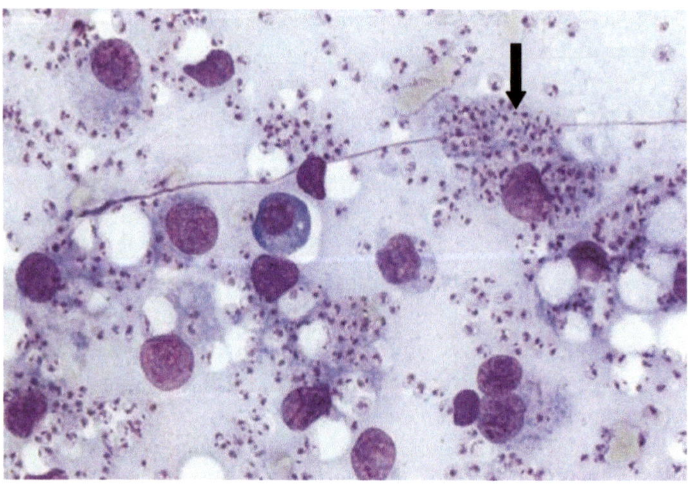

Fig. 4.13 Yeast form of Histoplasma capsulatum within macrophages. *Arrow* indicates the presence of capsule like material surrounding the yeast cell (Photograph courtesy: Prof. P.P. Gupta, Ex-Director of Veterinary Research, Punjab Agricultural University, Punjab, India)

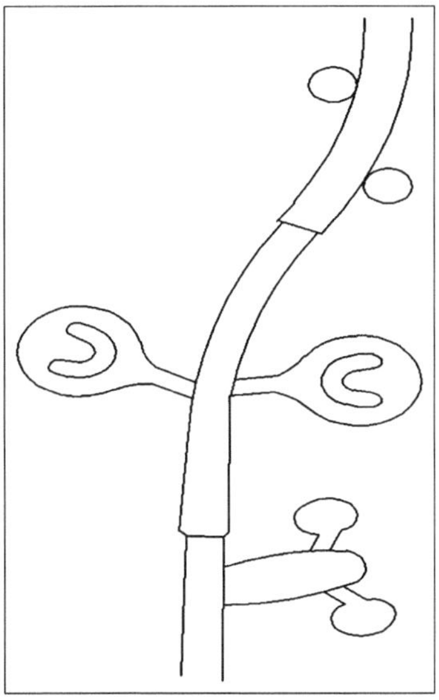

Fig. 4.14 Pear-shaped microconidia (schematic)

methenamine silver, periodic acid–Schiff (PAS), Giemsa and Gram stain, haematoxylin and eosin stain, whereas *H. capsulatum var. farciminosum* yeast cells appear as large double-contoured oval body in pus samples, measuring 2.5–3.5 μm × 3–4 μm. The cytoplasm is granular and sometimes budding is observed.

The mycelial form is characterised by the presence of aerial septate hyphae bearing microconidia (2–3 μm in diameter, Fig. 4.14) which is later converted into large chlamydospores (*tuberculate macroconidia*, 7–15 μm in diameter, Fig. 4.15) having small spine-like projections in their wall. However, the tuberculate macroconidia are not abundantly produced by *H. capsulatum* var. *farciminosum*. Instead other spores such as arthroconidia and blastoconidia are detected.

The carbohydrate is the major constituent (80 %) of the cell wall of both the morphotypes. The glucose (Glc), mannose (Man) and galactose (Gal) are the most abundant monosaccharide in the cell wall of *Histoplasma capsulatum*. The glucan (Glc) and N-acetylglucosamine persists as a polymer (β-1,4 linkage) known as 'chitin'. Two different glycosidic linkages (α and β) are present in the glucan structure of both yeast and mycelium. An α-glucan contains α-(1, 3)-glucosyl linear residues, while β-glucan consists of a linear β-(1, 3)-glucosyl-linked backbone with β-(1, 6)-glucosyl-linked side chains that vary in length and distribution. Especially in the yeast cells of *Histoplasma*, the chitin forms the inner layer with an outer fibrous covering made of α-glucan, whereas the mycelial form is predominantly made of β-glucan. Based on this chitin and α-glucan concentration, *H. capsulatum* is classified as chemotype I and II. The chemotype II strains consist of less chitin and more glucan than chemotype I. Further, concentration of α-glucan is associated with virulence of the strains. The virulent *H. capsulatum* strains may possess 1,000-fold more α-glucan than non-pathogenic strains. Sometimes *H. capsulatum* produces nonpathogenic strains lacking the α-glucan in their cell wall. They may produce an unusual morphology and can persist within the macrophages for prolonged period, observed in chronic animal and human infections. The α-glucan is found to be associated with multiplication of yeast cells within the macrophages, maintenance of yeast cells within phagolysosome and blocking of innate immune recognition of *H. capsulatum* by a particular receptor.

In addition the cell wall contains mannan, mannosylated proteins and low amount of lipids. The lipids are linked with saccharides and show structural heterogeneity. The glycosyl inositol phosphoryl ceramides present in the mycelial and yeast phases of *H. capsulatum* is required for fungal survival.

The conversion from mould phase to yeast phase is known as 'phenotypic switch' which is required for virulence as the yeast phase is protected from phagocytosis by the neutrophils, macrophages and monocytes due to its larger size to engulf. In addition, changes in chemical composition of the cell wall are also noted. In yeast phase, the α-glucan content of the cell wall is increased than the mycelial phase as discussed earlier and also the changes occurred in lipid composition of the membrane due to adaptation in higher temperature (37 °C). Some phase-specific virulence factors are expressed by *H. capsulatum* yeast cells such as CBP1, Yps 3 and Yps 21:E9. The sulphur

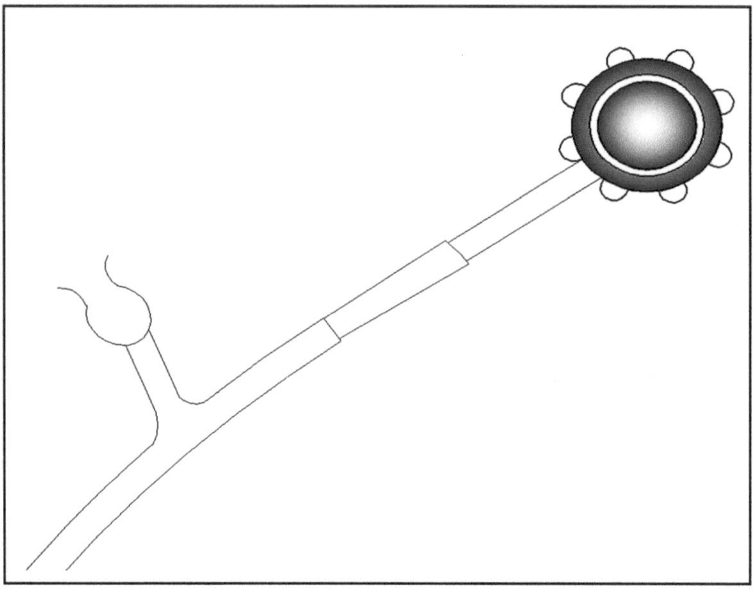

Fig. 4.15 Tuberculate macroconidia (schematic)

metabolism can influence the morphotype in *H. capsulatum*. Addition of sulfhydryl-reducing agent (dithiothreitol) fixes cells in the yeast phase independent of temperature, whereas the addition of sulfhydryl-oxidising agents (p-chloromercuriphenylsulfonic acid, PCMS) fixes cells in the mycelial phase independent of temperature. The enzyme hybrid histidine kinase (DRK1) regulates this phenotypic switch in *H. capsulatum*. It acts as a sensor in the change of environmental temperature and regulates the expression of the genes for yeast phase-specific virulence factors like CBP1, AGS1 and Yps-3. This phenotypic change is reversible and blocking the transition may prevent the disease progression.

4.7.2 Classification

The genus *Histoplasma* belongs to the phylum Ascomycota, class Euascomycetes, order Onygenales and family Onygenaceae like *Blastomyces* and *Coccidioides*. The genus possesses one major pathogenic species *H. capsulatum*, subdivided into three subspecies, i.e. *H. capsulatum* var. *capsulatum* and *H. capsulatum* var. *farciminosum* and *H. capsulatum* var. *duboisii*. *H. capsulatum* var. *farciminosum* was previously designated as a separate species (*H. farciminosum*) under the genus. Due to morphological similarities in both mycelial and yeast phases presently, it is assigned as a subspecies of *H. capsulatum*. Similarly, Vanbreusegham initially described *H. capsulatum* var. *duboisii* as a new species of the genus (*H. duboisii*) which was later reassigned within the *H. capsulatum* group. This *H. capsulatum* var. *duboisii* is more prevalent in Western and Central Africa and in the island of Madagascar.

H. capsulatum is also divided into phylogenetically distinct lineages/clades based on geographical distribution. The major clades are North American class 1 (Nam1), North American class 2 (Nam2), Eurasian, Latin American group A (LamA), Latin American group B (LamB), Australian, Netherlands (Indonesian) and African.

4.7.3 Susceptibility to Disinfectants

Histoplasma capsulatum var. *capsulatum* is susceptible to 1 % solution of sodium hypochlorite, 2 % phenol, 2 % gluteraldehyde and isopropyl alcohol and 3 % formalin. However, for decontamination of soil, only formalin is found effective. Among the antifungals, it is susceptible to amphotericin B and the azole group of antifungal drugs, including ketoconazole, itraconazole, fluconazole, posaconazole and voriconazole. The spores and yeast phase can be destroyed by exposure to more than 40 °C temperatures for an extended period of time, pH below 5 or above 10 and drying.

4.7.4 Natural Habitat

Histoplasma is a noncompetitive soil saprobe. It can grow in different soil types including hard clay and sandy loam. They are present within few centimetres from the soil surface where a microbiota is formed with bacteria, actinobacteria and other fungi. As they are noncompetitive saprobes, other soil microorganisms, which are active organic matter decomposers, favour their growth with availability of nutrient sources and contribute to the maintenance of soil structure, aeration and water content. Humid climate, temperature ranging from 25 to 35 °C and neutral to alkaline pH are required for their optimum growth. Sometimes darkness favours the sporulation and mycelial phase requires Ca^{++} for their growth. The bird droppings and bat guano can enrich their growth with nitrogen supplementation. The birds act as carrier of *Histoplasma*, whereas in bats (including insectivorous bats such as *Mormoops megalophylla*, *Myotis californicus*, *Pteronotus parnellii*) they cause intestinal infection. Especially the bat guano found in enclosed places, such as grottos, caves, mines and abandoned buildings, acts as a major source of infection. Most of the human outbreaks occur when they disturb the soil of old bird roosting site or they visit bat-inhabited caves for adventure or tour. The bat guano represents a complex habitat of mites, insects, bacteria and fungi. Among them the mites (*Sancassania*) are dependent on the nutrients provided by the mycelial phase of *H. capsulatum* to complete their life cycle. They are mycophagous and can be used as biological control agent against the fungi.

4.7.5 Genome

Histoplasma capsulatum strains are extremely diverse in terms of their number of chromosomes and genome size. For example, Downs strain has seven chromosomes, G186B has at least four, and G217B has at least three. The genome is haploid, and the genome size ranges between 23 Mbp (G186AS strain) to 32 Mbp (Downs strain), and the repetitive DNA content ranges between 0.5 and 8.0 %, respectively. The genome size is larger than ascomycetous yeast *Saccharomyces cerevisiae* (13 Mbp) and similar to those of the filamentous ascomycetes such as *Aspergillus nidulans* (26 Mbp), *Penicillium paxilli* (23 Mbp) and *Neurospora crassa* (43–45 Mbp). The nuclear guanine and cytosine (G + C) content is 45.4–49.8 %, with an observed mean of 47.3 %.

4.7.6 Isolation, Growth and Colony Characteristics

The mycelial phase of *H. capsulatum var. farciminosum* can be isolated in Sabouraud dextrose agar enriched with 2.5 % glycerol, mycobiotic agar, Hartley digest agar with 10 % horse serum, brain–heart infusion agar with 10 % horse blood and pleuropneumonia-like organism (PPLO) agar enriched with 2 % dextrose and 2.5 % glycerol. The addition of antibiotics to the media is recommended to prevent the growth of contaminant bacteria. The recommended antibiotics include cycloheximide (0.5 g/L) and chloramphenicol (0.5 g/L). The plates are kept at room temperature (25–30 °C) for 2–8 weeks.

For conversion into yeast phase, subculture of mycelium can be performed in brain–heart infusion agar containing 5 % horse blood or in Pine's medium at 35–37 °C with 5 % CO_2 tension and high humidity. The complete conversion becomes possible after 4–5 repeated subcultures into fresh medium every 7–8 days depending upon the strain. The conversion is also possible by inoculation of the mycelium into cell monolayer (HeLa) or in experimental mice.

The mycelial phase appears as dry, grey-white, granular, wrinkled mycelium. The colonies become brown with aging. The colonies chiefly contain sterile hyphae with rare occurrence of chlamydospores, arthospores and blastopsores.

The yeast colonies are small, grey, flaky composed of yeast cells and a few hyphae. *H. capsulatum* var. *capsulatum* yeast cells produce flat, raised, wrinkled, and white to greyish brown colonies which often revert into its stable mycelial phase.

4.7.7 Antigenic Characteristics

Cell wall mannans (Man-containing polysaccharides) and mannosylated proteins are major antigens of *H. capsulatum* var. *capsulatum* which help in host tissue adherence. Galactomannan–protein complexes from the mycelial phase cell wall of *H. capsulatum* can induce delayed-type hypersensitivity in guinea pigs and inhibit macrophage migration factor release. The filtrates of the mycelial culture (Histoplasmin) are enriched with this protein complex. The galactomannan complexes may also protect the organism against its own serine-thiol protease, an enzyme associated with pathogen dissemination through the extracellular matrix. In addition, heat shock protein (HSP60) of *H. capsulatum* var. *capsulatum* is considered as an immunodominant antigen; vaccination of mice with this protein protected them from lethal and sublethal histoplasmosis due to its stimulatory nature of CD4+ T cells. The M antigen (catalase B) is a major diagnostic antigen as it can induce both the cellular and humoral immune responses.

Antigenically no difference is observed between *H. capsulatum* var. *capsulatum* and *H. capsulatum* var. *farciminosum.*

4.7.8 Virulence Factors

The virulence factors possessed by *H. capsulatum* are described in Table 4.18.

4.7.9 Transmission

H. capsulatum var. *farciminosum* is transmitted into susceptible hosts (horses) by wound infection through the nasal or ocular secretions, contaminated saddlery, grooming utensils, soil or harnesses, feeding and watering utensils, etc. The flies and ticks can also help in transmission of the infection. Especially the conjunctival form of the disease is transmitted by flies (*Musca* or *Stomoxys*). Rarely inhalation, ingestion of contaminated feed act as possible ways.

H. capsulatum var. *capsulatum* is mostly transmitted by inhalation, ingestion and rarely by wound contamination. In human, males are found more susceptible especially those exposed with the area where the birds roost or the bats inhabit. In such areas, infectious particles can exceed 10^5 particles per gram of soil. The people working as farmers, gardeners, poultry farmers and cave explorers are at high risk of contamination, and their occupations can be designated as 'histo-hazard' jobs. Recent evidence suggests that exposure to a bamboo bonfire may also transmit the infection. The heating increases the transmission efficiency of the spores present in the bamboo, contaminated with bird faeces.

Inhalation is the probable route of entry for *H. capsulatum* var. *duboisii* in human. Rarely transmission through transcutaneous route is found.

4.7.10 Pathogenesis

4.7.10.1 *H. capsulatum* var. *capsulatum*

H. capsulatum var. *capsulatum* is soil-based fungi and they do not require human or animals to propagate. So mammals act as incidental host in their life cycle. After inhalation of spores, transition into the yeast phase occurs in higher body temperature of the host, required for expression of virulence genes and pathogenicity.

The cellular entry of *H. capsulatum* var. *capsulatum* yeast phase occurs either by

Table 4.18 Virulence factors of *H. capsulatum*

Virulence factors	Location in fungal cell	Function
α (1, 3)-glucan	Outermost layer of yeast cell wall	(i) It conceals immunostimulatory β-glucan layer of the cell wall from detection by host phagocytic cells. It blocks the recognition of β-glucan through its Dectin1 receptor. So it acts as 'decoy ligand' for Dectin1 receptor (ii) It inhibits the production of the pro-inflammatory cytokine (TNFα) by phagocytes
Heat shock protein (HSP60)	Cell wall of both yeast and mycelium	(i) It acts as ligand for CR3 (complement receptor 3) in the macrophages. Specifically it binds with the CD18 chain of CR3. Their attachment helps the fungi to enter the macrophage and initiate the intracellular phase. This kind of binding of carbohydrate ligand (Man, GlcNAc and Glc) with CR3 repeals the release of toxic oxygen metabolites. So it acts as safe portal for entry into the macrophages (ii) The binding of HSP60 with CR3 downregulates host IL12 production, required for protection against fungal infection (iii) Chaperoning of proteins (associated with oxidative stress response) to help their expression in fungal cell surface
Calcium-binding protein (CBP1)	Secreted by yeast cells	It helps in survival of yeast cells in macrophages in vitro, by acquiring calcium from the environment
Yeast phase-specific protein 3 (Yps3)	Yeast cell wall localised protein and occasionally secreted (it is localised with the help of epidermal growth factor (EGF)-like domain that can attach with chitin of the cell wall)	Absence of this protein in the cell wall hampers the growth of the fungi in the lung and spleen. Exact role in pathogenesis is uncertain
(i) Catalase B (M antigen; 70–94 KDa) (ii) Catalase P (57 KDa) (iii) Catalase A	Constitutively expressed protein of yeast cells Monofunctional peroxisomal catalase, constitutively expressed protein of yeast cells Large subunit bifunctional enzyme, induced upon H_2O_2 stress	It prevents the lethal action of hydrogen peroxide produced during oxidative burst defence mechanism within the host phagocytes
Beta-glucosidases (H-antigen)	Secreted protein of yeast cells	It helps in remodelling of the cell wall to acquire more nutrients
melanin	Cell wall of the yeast and conidia	It protects the fungi from host-derived free radicals, microbicidal peptides and antifungal drugs
Dimerum acid, coprogen B and fusarinine	Secreted	They act as siderophores in vitro under iron-limited condition. It scavenges for ferric iron (+3 redox state). Exact mechanism of iron acquisition is unexplored
(a) Reduced glutathione-dependent enzymatic reductase (b) Low-molecular-weight nonenzymatic reductants	Secreted	They can reduce Fe^{+++} to Fe^{++} and help in transport of ferrous iron which is a way for acquisition of iron from inorganic or organic salts or siderophores, binding with the ferric iron
Histone 2B (H2B)	Cell surface of yeasts	Exact role in virulence is uncertain, neutralisation of H2B with monoclonal antibody reduced fungal burdens, decreased pulmonary inflammatory damage and prolonged experimental mice survival
70 KDa antigen	Cell surface of yeast cells	Role in virulence is uncertain

'host-directed way' or 'fungi-directed way'. The host-directed way requires opsonisation of the yeast cells with antibody or complement to enter the macrophages. In fungi-directed way, they can directly bind the heterodimeric β2 or CD18 integrins on the surface of host mononuclear cells, including LFA-1 (with αL or CD11a), Mac-1/CR3 (with αM or CD11b) and P150, 95/CR4 (with αX or CD11c). The heat shock protein (HSP60) acts as major fungal ligand for this kind of attachment. In addition to the macrophages, the yeast phase of fungi can also enter the dendritic cells through the interaction between fungal adhesin (cyclophilin) and fibronectin receptor (very late antigen 5).

Within the host cells after internalisation, the yeasts enter into the phagosomes which fuse with lysosomes to form phago-lysosome. *H. capsulatum* var. *capsulatum* follows different strategies to survive within the phago-lysosome. They can increase the pH of the phago-lysosome with the help of urease and release of ammonia or bicarbonate, although it is not established whether they inhibit or reverse the lowering of pH. It also prohibits the accumulation of vacuolar ATPase/proton pump required for maintenance of H^+. Thus, they elevate the pH of the phago-lysosome up to 6.5, optimum for iron acquisition from transferrin with the help of siderophores and lowering the action of lysosomal hydrolase. Calcium-binding protein released by the yeasts also lowers the lysosomal enzyme capacity by chelation of calcium. Other virulence factors such as catalase B, beta-glucosidase and melanin help in the survival of the yeast cells within macrophages as described in Table 4.18. The infected macrophages enter the lymph nodes and spread subsequently through haematogenous route. In the absence of a protective cellular immunity, the disease with dissemination may occur. Sometimes the yeasts remain dormant in the host throughout their life and rejuvenate to produce disease during impairment of host immunity.

4.7.10.2 *H. capsulatum* var. *farciminosum*

After transmission into the host, the exact mode of dissemination especially at the cellular level is unexplored. It causes suppurating lesions in the skin of limbs and neck associated with the lymphatics followed by the development of ulcer and enlargement of the lymphatics with tortuous cord formation. In mucosal form of the disease, lesions are observed in the nasal passage, pharynx, larynx, trachea, lung and genitalia with possibility of venereal transmission. In ocular form, conjunctivitis, keratitis, papule formation in conjunctiva or nictitating membrane is found.

4.7.11 Disease Produced

The major animal and human diseases produced by different subspecies of *H. capsulatum* are enlisted in Table 4.19.

4.7.12 Immunity

4.7.12.1 *H. capsulatum* var. *capsulatum*

The cell-mediated immunity (Th1 response) can protect the host as the fungi (*H. capsulatum* var. *capsulatum*) are intracellular in nature. The cell-mediated immunity develops 10 days post-infection, primarily with the production of IL12 followed by IFN γ which is crucial for protection. The macrophages are the primary target cells of the yeasts where they can replicate. However, for induction of protective cellular immunity, the fungicidal activity of the macrophages should be enhanced. Phagocytosis of the yeasts with dendritic cells effectively clears them as they are fungicidal. Azurophilic granules of neutrophils have fungistatic activity in vitro.

The IL12 is chiefly produced by dendritic cells and neutrophils as the production capacity of macrophages are downregulated by the fungi. The IL12 stimulates the IFNγ production by CD4+ T cells, CD8+ T cells and NK cells. Among them CD4+ T cells are the predominant producers of IFN γ in *H. capsulatum* infection. The IFN γ can enhance the fungicidal activity of macrophages with the production of nitric oxide and can limit the accessibility of iron and zinc to the fungi required for their intracellular replication. The pro-inflammatory cytokine (TNFα)

Table 4.19 Major diseases of animal and human caused by *Histoplasma*

Fungi	Host	Disease
H. capsulatum var. *farciminosum*	Horse, mule, donkey, rarely in camel, dog and cattle. It is transmissible to human (zoonotic)	*Epizootic lymphangitis (African glanders/pseudo-glanders/pseudo-farcy)*: It is characterised by suppurating, ulcerative pyogranulomatous dermatitis and lymphangitis especially in the limbs, neck and chest. Nodular lesions appear in the skin, subcutaneous tissues and lymphatics. In other forms of the disease, conjunctivitis, keratitis, and multifocal pneumonia are also observed. The disease is endemic in tropical countries such as North, East and Northeast Africa and some parts of Asia such as India, Pakistan and Japan. It occurs sporadically in other parts of the world such as France, Russia, Egypt and Italy
H. capsulatum var. *capsulatum*	Human	Two clinical forms, i.e. self-limiting mild respiratory tract infection and disseminated forms, are noted. The later form is found life-threatening in elder or immunocompromised patients such as AIDS patients or recipients of organ transplants. The fungus is endemic worldwide, but there are regions with high incidences of infection, such as areas along the Ohio and Mississippi River Valleys in the United States and in Rio de Janeiro State in South eastern Brazil
	Dog	Three clinical forms, i.e. pulmonary, disseminated and clinically inapparent form, are found. Appearance of clinical signs depends on organ involvement. In general the signs include coughing, dyspnoea, anorexia, ascites and ulceration of oral and nasal mucosa, fever and diarrhoea
	Cat	Respiratory disease, fluctuating fever, weight loss, lethargy
H. capsulatum var. *duboisii*	Human	*African histoplasmosis*: This form is prevalent in western and central Africa and in the island of Madagascar. The disease involves the skin, subcutaneous tissue, lymph node, bone and rarely the lung or other organs

also plays an important role in protection. It can also stimulate the fungicidal activity of macrophages with the production of nitric oxide and indirectly organises cellular migration by controlling chemokine production and adjusts the emergence of regulatory T cells (Tregs). Tregs in TNFα-neutralised mice dampen the protective immune response in an IL10-dependent manner. Other Th1-associated cytokines such as granulocyte-macrophage colony-stimulating factor (GM-CSF) stimulates macrophage fungistatic activity against *H. capsulatum* yeast.

Recently the role of Th17 response in *H. capsulatum* infection is also explored. In pulmonary histoplasmosis, IL17, IL6 and IL23 gradually increase the first week of infection and declines thereafter. CD4+ and CD8+ T cells expressing CD25 are the predominant source of IL17 during infection. IL17 can enhance the entry of inflammatory cells into the lungs; however, it directly has minimal role in protection, whereas in the absence of IL12 (Th1 response), IL23 (Th17 response) plays a vital role in protection.

In contrast, Th2 type of immunity generated by IL4 delays the clearance of *H. capsulatum* from the lungs and spleen. IL4 enhances the intracellular fungal growth by reducing nitric oxide production and increasing the availability of iron, zinc and calcium.

The antibody response to *Histoplasma capsulatum* infection is characterised with an increase in IgM by 2 weeks, followed by rising titres of IgA and IgG. The IgG fraction contains complement-fixing and precipitating antibodies. However, the protective role of humoral immunity is uncertain. Experimentally no significant difference was observed between B cell-deficient and the wild-type mice in primary and secondary histoplasmosis.

4.7.12.2 *H. capsulatum* var. *farciminosum*

Cellular immunity plays a protective role in epizootic lymphangitis. The antibodies develop after appearance of the clinical signs. The infected animals produce a delayed hypersensitivity reaction after inoculation of mycelial extract (histofarcin).

4.7.13 Diagnosis

4.7.13.1 Clinical Specimens

For detection of *H. capsulatum* var. *farciminosum*, the exudate or biopsy samples can be collected from unruptured nodules with the help of fine needle. The material should be kept in sterile peptone water with antibiotics and transported to the laboratory in a flask with ice and proper label. The swabs of the material should be prepared for direct examination. For histopathology, sections of lesion material should be placed in 10 % neutral-buffered formalin.

For detection of *H. capsulatum* var. *capsulatum*, urine, blood, bronchoalveolar lavage (pneumonia patients), cerebrospinal fluid (meningitis patients) and bone marrow aspirates can be collected from human patients. Pulmonary aspirate, gastric washings, peripheral blood (febrile condition), faeces, sternal bone marrow, biopsied liver, spleen or lymph node can be collected from dogs.

4.7.13.2 Laboratory Examination

1. *Direct examination*: The wet mount or gram-stained smear is prepared with the swab sample and observed under microscope for detection of typical *H. capsulatum* var. *farciminosum* yeast cells, which will appear as gram positive, pleomorphic or pear shaped, double contoured, approximately 2–5 μm in diameter. They are intracellular within macrophages or neutrophils or occasionally they are extracellular. Budding may be visible from the pointed end of the yeast cells. The electron microscopy can be applied for detection of *H. capsulatum* var. *farciminosum* with minute cellular details in the skin biopsy samples.

   Direct examination provides best result for detection of *H. capsulatum* var. *capsulatum* in pulmonary histoplasmosis patients. In South America, Tzanck cytodiagnosis technique provides a useful and rapid diagnosis method for AIDS-associated histoplasmosis, due to the high frequency of mucocutaneous lesions in these patients.
2. *Isolation and identification*: Isolation of *H. capsulatum* var. *capsulatum* from bone marrow aspirates and blood samples suffering from disseminated histoplasmosis is an important diagnostic tool. In blood culture, the lysis-centrifugation technique should be employed, since it is sevenfolds more sensitive than the conventional techniques used for clinical sample preparation. Media and incubation condition as described earlier will serve the purpose.
3. *Animal inoculation test*: Laboratory animals such as mice, guinea pig and rabbits are used for experimental inoculation of *H. capsulatum* var. *farciminosum*. Impression smear of the liver and spleen from the inoculated animals should show the yeast cells 2–4 weeks post inoculation in positive cases.
4. *Detection of antigen*: ELISA and fluorescent antibody technique (FAT) are successfully used for the detection of H/M antigen of *H. capsulatum* var. *farciminosum* from cultured mycelium. Enzyme immunoassays are also used for the detection of *H. capsulatum* var. *capsulatum* circulating polysaccharide antigens in the immunocompromised human patients with a disseminated form.
5. *Serological tests*:
   (a) Passive haemagglutination, immunodiffusion, complement fixation test (CFT), radioimmunoassay (RIA) and ELISA were developed to detect *H. capsulatum* var. *capsulatum* antibodies from human and animal clinical samples.

   Immunodiffusion test detects the presence of two types of precipitin bands, i.e. M and H bands. The M band develops in both acute and chronic infection and may be detected for years even after recovery, whereas the H band is rare and, if found, is associated with chronic or severe acute infection. The assay is 80 % sensitive but more specific than CFT. It gives positive results in the majority of progressive pulmonary and disseminated forms, and its titres are related to the fungal burden, so they decrease with the successful treatment. Immunodiffusion is not very useful in HIV-positive patients with histoplasmosis, since it gives positive results in only 35 % of mycologically proved cases of histoplasmosis.

The CFT uses two types of antigens, i.e. yeast and mycelial (histoplasmin). The diagnosis is based on fourfold rise in CF antibody titre; a single titre of ≥1:32 is suggestive but not diagnostic. The CF antibody persists for more than 1 year. The test is cross-reactive with other fungal infection and granulomatous diseases such as tuberculosis and sarcoidosis.

Histoplasma antibody was detected in the urine of 89 % of human patients with disseminated form of histoplasmosis and 37 % of human patients with other forms of histoplasmosis by RIA and ELISA methods. The specificity was noted as 99 %. However, false-positive result was noted in transplant patients receiving rabbit antithymocyte globulin. The serological response is strongly detectable in pulmonary histoplasmosis than disseminated and asymptomatic forms due to lower intensity of exposure resulting in reduced fungal burden and antigenic simulation or due to immunosuppression.

(b) Skin hypersensitivity test (histofarcin test)

The mycelial form of *H. capsulatum* var. *farciminosum* is grown on polystyrene discs floating PPLO media containing 2 % glucose and 2.5 % glycerine at 23–25 °C for 4 months. The fungus-free culture filtrate is mixed with acetone (2:1) and held at 4 °C for 48 h. The supernatant is decanted and the acetone is allowed to evaporate. The precipitate is suspended to 1/10 original volume in distilled water. The animals are inoculated intradermally with 0.1 ml of antigen in the neck. The inoculation site is examined for the presence of swelling at 24, 48 and 72 h postinjection. In positive case, an indurated and elevated area at the injection site will be observed.

6. *Molecular biology*: A PCR assay was developed for detection of *H. capsulatum* var. *capsulatum* from both the clinical samples and culture using the primers for the gene encoding M antigen. This assay had a sensitivity and specificity of 100 %.
   Real-time PCR showed promising result by detection of *H. capsulatum* var. *capsulatum* from tissue biopsies and bronchoalveolar lavage fluid from human patients who had documented histoplasmosis.

4.7.14 Treatment

In human suffering from pulmonary and disseminated form of histoplasmosis, liposomal amphotericin B (for 2–3 weeks) followed by itraconazole can be administered. In chronic pulmonary histoplasmosis, itraconazole and ketoconazole are effective. The course of treatment is 3 months to 1 year depending on the severity. Due to prolonged therapy, careful monitoring for nephrotoxicity (amphotericin B) and drug interaction (itraconazole) is required. Itraconazole is also used as prophylactic drug in patients of endemic zone having less CD4+ T cells. In AIDS patients, bioavailability of itraconazole is decreased. Oral posaconazole and voriconazole are currently considered as alternative treatments for non-severe disseminated histoplasmosis. Fluconazole is less effective but can be used in the patients who are allergic to itraconazole.

In horses suffering from epizootic lymphangitis, surgical removal of lesion along with amphotericin B administration is recommended.

In dog and cats, drug of choice is itraconazole due to low level of toxicity. Other drugs such as ketoconazole and fluconazole are also effective. Pulmonary form is curable with proper treatment. The prognosis of disseminated form is grave.

4.8 *Rhinosporidium*

In 1900, Seeber first published the report of rhinosporidiosis which occurred in an Argentinean youth suffering from laboured breathing due to nasal polyp. Seeber although noted that Malbram (1892) was the first to diagnose the disease in human in Argentina which was unpublished. Seeber described the similarities of the organism with *Coccidioides immitis*. Accordingly, Robert Wernicke, advisor of Seeber, named the organism as *Coccidioides seeberia* or *Coccidium seeberi* in 1900. In 1903, O'Kineley first described the histological changes caused by the organism which were isolated from India. Minchin and Fantham (1905) further studied O'Kineley's Indian isolates and named the organism as *Rhinosporidium kinealyi*. The earliest evidence of animal infection was noted by Theiler (1906) in South Africa in a horse. Later, similar organism from the horses was reported by ZSchokke (1913) in South Africa, who also named them as *Rhinosporidium equi*. Ashworth (1923) revised all the nomenclature proposals and finally concluded *Rhinosporidium seeberi* as the name of the pathogen. Thus, the different names such as *Rhinosporidium ayyari* (isolated from buffalo), *Rhinosporidium hylarum* and *Rhinosporidium amazonicum* are currently known as *R. seeberi*.

In India, animal rhinosporidiosis was reported by several workers such as Ayyar (1927) and Rao (1938, 1951). Ayyar (1927) reported for the first time about the occurrence of rhinosporidiosis in cattle in India.

4.8.1 Morphology and Life Cycle

The morphology and life cycle study of *Rhinosporidium seeberi* is based on its appearance in the tissue section, as the organism cannot be grown in vitro. Seeber (1900) first described the life cycle in the infected tissues. Primary juvenile sporangium (trophocyte) is a thin-walled structure with a single nucleus, nucleolus and granular cytoplasm containing mitochondria. The cytoplasm also contains endoplasmic reticulum, lipid-rich globules and vacuoles. The cell wall is unilamellar in nature with a fibrillar capsule of electron-dense material. This juvenile sporangium undergoes the sequential morphological changes to produce immature, intermediate, early mature and mature sporangia. The intermediate sporangia (70–150 μm) contain thickened bilamellar cell wall made of chitin (thinner electron-dense outer wall and thicker translucent inner wall), several nuclei, several flat cristae mitochondria and an early pore. The cytoplasm condenses around each of the nucleus to develop endospores (4–10 μm in diameter), and the sporangium with such endospores is known as early mature sporangia. The mature sporangium is a large (up to 400 μm) spherical body containing hundreds of mature and immature endospores and a pore or operculum. It is enclosed within a thin cell wall with three prominent inner layers. The mature endospores are located near the centre and towards the pore, whereas the immature endospores are located at the opposite pole of the sporangium (germinative zone) (Fig. 4.16).

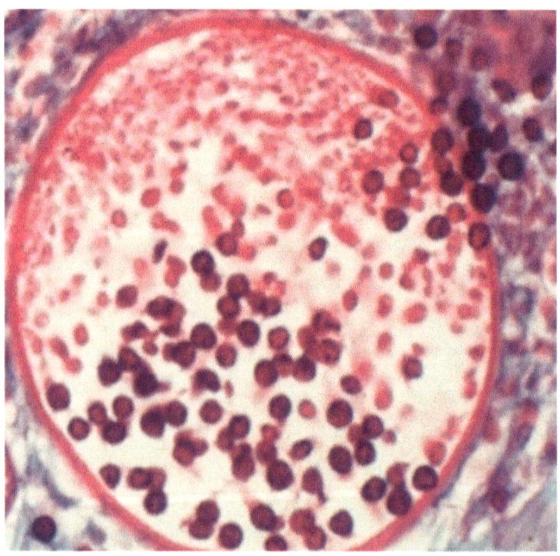

Fig. 4.16 Matured sporangium of *Rhinosporidium seeberi* in a nasal polyp stained with PAS (Photograph courtesy: Prof. P.P. Gupta, Ex-Director of Veterinary Research, Punjab Agricultural University, Punjab, India)

The mature endospores contain a bilaminated thin cell wall (1–3 μm), 15–20 spherical (1–3 μm in diameter) electron-dense bodies (EDB), nucleus with nucleolus, Golgi apparatus, endoplasmic reticulum and mitochondria. The electron-dense bodies either act as nutrient reservoir or precursor of endosporulated sporangia and they are known as 'spore-morula'.

The mature endospores are released from the sporangium through the pore. There are two explanations of the endospore release mechanism. In contact with watery material, lytic enzymes are accumulated at the pore site of the mature sporangia. The inner osmotic pressure of the mature sporangia is also increased which facilitates the release of endospores through the pore. Another explanation states that due to high antigenicity of the sporangia, anti-sporangia antibodies are developed by the host defence system. The immune complexes are produced by the interaction of the antigen and antibody. The complexes aid in expulsion of endospores by producing external pressure during expansion of the sporangia through increased water intake.

After release, the endospores increases in size (10–70 μm) and lose all the EDBs to become a juvenile sporangium and the cycle is repeated. The life cycle of *Rhinosporidium seeberi* is shown in Fig. 4.17.

4.8.2 Classification

The taxonomical status of *Rhinosporidium seeberi* is controversial because it could not be isolated in artificial culture. Previously, it was considered as a cyanobacterium under the genus *Microcystis* probably due to morphological resemblance between coccal-shaped *Microcystis* nanocytes and spherical endospores of *Rhinosporidium*. However, 16S rRNA sequence analysis revealed only 79–86 % similarity with *Microcystis*. The ultrastructural studies revealed that the nanocytes do not possess a true cell wall and are held together with an amorphous matrix, whereas the *R. seeberi* endospores are surrounded with a thin cell wall.

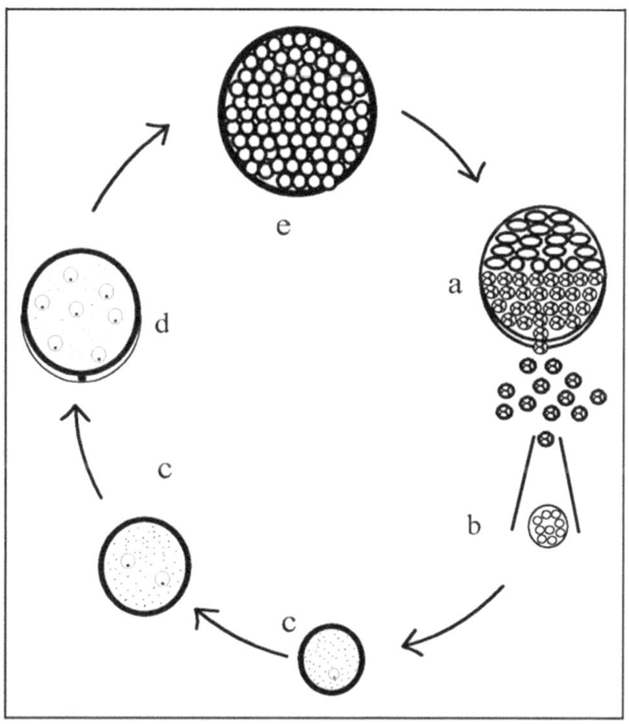

Fig. 4.17 Life cycle of *Rhinosporidium seeberi* (schematic); *a* release of endospores from matured sporangium, *b* released endospores become large (10–70 μm), *c* formation of juvenile sporangium possessing single or double nuclei with nucleolus and thick cell wall, *d* formation of intermediate sporangium (70–150 μm), *e* early matured sporangium (<150 μm)

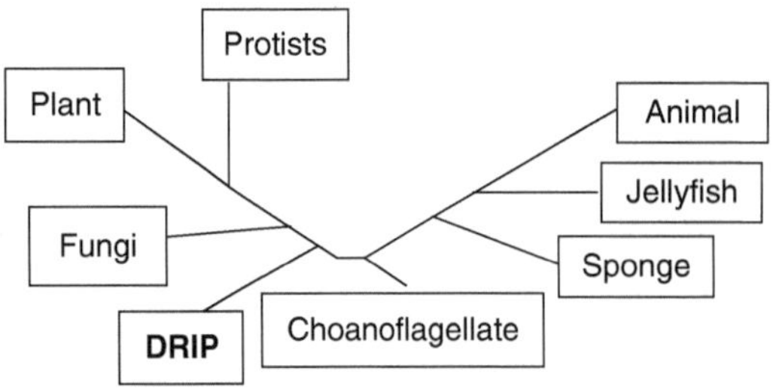

Fig. 4.18 Phylogenetic position of DRIP group

Further studies revealed its eukaryotic nature and claimed its morphological similarities with Chytridiales group of fungi. However, comparative studies by other workers showed certain morphological dissimilarities especially in the cell wall structure between *R. seeberi* and prototype member of Chytridiales. The most important dissimilarity is the compulsive intracellular nature of Chytridiales which is not possible in *R. seeberi* due to their larger size.

Later, phylogenetic and taxonomic analysis placed it under the new cluster of aquatic pathogens, known as the DRIP group (Dermocystidium, rosette agent which is presently known as *S. destruens*, *Ichthyophonus* and *Psorospermium*). The DRIP group was renamed as Mesomycetozoa class which is located at the divergence between animals and fungi (Fig. 4.18). The class is composed of two orders such as Dermocystida and Ichthyophonida. The morphological and phylogenetic analysis of *R. seeberi* revealed its similarity with the members of Dermocystida. The presence of synchronised nuclear division with the formation of endospores in the matured stage confirms its association with Mesomycetozoa class.

4.8.3 Susceptibility to Disinfectants

Experimentally metabolic inactivation with or without altered structural integrity of *R. seeberi* was observed with hydrogen peroxide, glutaraldehyde, chloroxylenol, chlorhexidine, cetrimide, thimerosal, 70 % ethanol, 10 % formalin, povidone–iodine, sodium azide and silver nitrate. The total period of exposure of endospores to the studied disinfectants was for 7 min which produced metabolic inactivation.

4.8.4 Natural Habitat and Distribution

Aquatic environment is detected as major ecological niche of *R. seeberi* because most of the patients are observed in close proximity with same water source in a locality such as lake, pond, etc. However, the fungi are also present in the dry areas of North India, Middle East countries and Argentina probably through the formation of a resistant structure.

The fungus is reported from more than 70 countries with highest incidence (95 %) in India and Srilanka which are considered as endemic zones. India and Srilanka is followed by South America, Brazil, Argentina and Mexico and a few reports are available from Nepal, Columbia, Venezuela, United States, Uganda, Madagascar, Ghana, Iran, Russia, Europe and Southeast Asia. Most of the cases in human and animals such as cattle, buffalo, goat, dog, cat, mule, duck, geese and water fowl are sporadic. The sole human and swan outbreak of ocular and nasal rhinosporiodiosis was reported so far from Serbia and Florida, United States, respectively.

4.8.5 Genome

The juvenile sporangium of *R. seeberi* contains four chromosomes. Other genome characterisation of *R. seeberi* is poorly known because most of the sequences deposited in the genbank (NCBI) are rDNA sequences rather than the protein-coding sequences. Further no information is also available about the genome of closely related mesomycetozoeans.

4.8.6 Isolation, Growth and Colony Characteristics

Since the time of discovery, attempts have made to isolate *R. seeberi* in artificial and live cell culture without any success. Although a few workers have claimed about isolation of *R. seeberi* in artificial culture and in human rectal epitheloid cell line, further studies failed to support their statement. The confirmational studies revealed that the isolates were environmental contaminant. In crushed rhinosporidial tissues, formation of endospores and sporangium was noticed when kept at low temperature or room temperature. However, the sustained propagation with repetition of the whole life cycle was not found in any culture system.

4.8.7 Antigenic Characteristics

In the mature sporangium of *R. seeberi*, an immunodominant antigen is detected which is absent in juvenile or early mature sporangium and the endospores. The antigen is located in the inner electron lucent cytoplasmic layers of the mature sporangium. Another study revealed the presence of a circulating antigen in the sera of the patients infected with *R. seeberi*. However, further characterisation is needed to detect their chemical structure.

4.8.8 Transmission

In human, nasal form is the most common form of rhinosporidiosis which occurs due to frequent bathing or working in stagnant fresh water. The skin injury is the most frequent predisposing factor. It is commonly observed in river-sand workers in India and Srilanka. The sand particles cause skin abrasions through which the infection can occur during direct contact with the contaminated river water. However, in arid zones such as in North India and Middle East countries, ocular form predominates and the resistant endospores are considered as major infectious agent. The endospores are transmitted through air especially during sand storms.

In animals, the infection takes place through traumatised skin from the contaminated water bodies. The hunting or street dogs are in increased risk due to more exposure to these factors. There is no clear evidence of transmission between human, animals, human to animals and vice versa.

4.8.9 Pathogenesis

The primary lesion develops in the nasal passage of the host after transmission of the infection. The rhinosporidial lesions are polypoidal, less than 3 cm in diameter, pedunculated or sessile, granular, multilobed and red in colour due to marked vascularity. Due to pronounced vascularity, the lesion bleeds easily. The surface of the lesion contains yellowish small spots underlying mature sporangia. The polypoidal lesions developed in the face and trunk are converted into verrucous warts.

The dissemination of infection may occur through the blood circulation into the distant viscera. The local spread may occur into the adjacent area of the polypoidal lesions by spillage of the endospores during trauma or surgery of the polyp. It is known as 'autoinoculation'.

The disseminated lesions have been detected in the conjunctival sac, ear, vagina, limbs, trunk and other external part of the body where it appears as subcutaneous lump with intact skin. The lesions are ulcerated growths resembling sarcomas and carcinomas. The involvement of the limbs often causes the destruction of underlying bones. The dissemination of the infection into the central nervous system has fatal consequence.

4.8.10 Disease Produced

There are four clinical forms of rhinosporidiosis detected in human patients. In nasal form, the lesions (polyp) develop at the anterior nares, nasal cavity, septum and floor. The nasopharynx, larynx and soft palate are sometimes affected. The formation of polyp is painless until it becomes large enough to obstruct the nasal passage or to put pressure on the nerve and vasculature endings. In ocular form, the lesions develop in the bulbar and palpebral conjunctiva.

The ocular lesion begins with a sessile growth which later becomes pedunculated, degenerating to friable. If the lesion becomes large, there is redness of eyes, tearing, photophobia, lid eversion and conjunctivitis. The cutaneous form is characterised by formation of papule initially which later becomes wart-like growth with a crenulated surface. In the disseminated form, the sporangia are detected from the bone, liver, spleen, lung and brain.

In dogs, the clinical signs such as visible mass within the nares, wheezing, sneezing, unilateral purulent nasal discharge and epistaxis are noticed.

4.8.11 Immunity

The cell-mediated immune (CMI) response (Th1 mediated) is generated against *R. seeberi* along with suppressor reaction. The major cell population involved in CMI response are macrophages (CD68), NK cells (CD57+), numerous CD8+ T lymphocytes and less numbers of CD4+ T lymphocytes. In chronic and disseminated infection, the immune response is converted into Th2-mediated response. The antibodies are generated without any protective role against infective endospores. The antibodies are detected to bind the soluble antigen of *R. seeberi* which is released by sonication of the sporangia in vitro. Probably it acts as an immune evasion mechanism which protects the endospores from the antibodies in vivo.

4.8.12 Diagnosis

4.8.12.1 Clinical Specimens

The clinical specimens for rhinosporidiosis include nasal scrape, fine needle aspirates from the lesion and tissue biopsy specimens. The nasal scrape is preferred over the fine needle aspirates because the lesions bleed easily.

4.8.12.2 Laboratory Examination

1. *Direct examination*: The KOH mount can be prepared from the clinical samples such as macerated polyp to detect the sporangia (350–400 μm in diameter) with endospores or the empty sporangia and free endospores in positive samples. It may produce confusion with the spherules of *Coccidioides immitis* which are smaller (60–100 μm diameter; Fig. 4.12) than the mature sporangia of *R. seeberi*.
2. *Histopathology*: The histological inspection of the lesion detected hyperplastic epithelium with sporangia containing endospores below the surface. The area is highly vascularised and surrounded by a connective tissue layer. Different immune cells such as polymorphonuclear cells, lymphocytes, plasma cells, giant cells and histiocytes infiltrate into the area. Different stages of *R. seeberi* can be observed with special fungus stains such as the Gomori methenamine silver, Gridley's, periodic acid–Schiff stains and haematoxylin–eosin stain.
3. *Molecular biology*: The polymerase chain reaction can detect *R. seeberi* by amplification of the 18S rRNA gene directly from the infected tissues.

4.8.13 Treatment

Spontaneous regression of the nasal growth is observed in human and animals which is rare in occurrence. So, surgical intervention or treatment with drugs is required. Radiotherapy has no effect in curing of the rhinosporidial polyp. The surgery by hot or cold snare techniques is the treatment of choice; however, recurrence of polyp may occur sometimes (10 %).

In animals, berenil (diminazine aceturate) and imizol (imidocarb dipropionate) can be used. The human and animal drugs such as amphotericin B, dapsone, ketoconazole, trimethoprim–-sulphadiazine and sodium stibogluconate showed effective in vitro anti-rhinosporidial activity. However, in vivo trial report is available with amphotericin B which was effective topically. Instead, dapsone may be a better choice for treatment with a dosage of 100 mg/kg body weight orally for 6 months to 1 year.

4.9 *Rhizopus*

The genus *Rhizopus* was first established in 1820 by Ehrenberg, and the fungus was chiefly known for the production of fermentation products like ethanol, lactic acid, fumaric acid and malic acid. The pathogenic potentiality was recognised later when the human infection (rhinocerebral zygomycosis) due to Zygomycetes was detected in Germany by Platauf (1885). He described the infection as 'mycosis mucorina'. The sketches drawn by Paltauf to describe the etiological agent were more similar with *Rhizopus* than *Mucor*. In 1943, similar kinds of human cases, affecting meninges and central nervous system, were reported from patients with diabetes mellitus. Bauer et al. (1955) first isolated the major etiologic agent of rhinocerebral zygomycosis, i.e. *Rhizopus arrhizus* (*R. oryzae*).

Baker (1957) coined the term 'mucormycosis' to describe Zygomycetes infection. Clark (1968) restricted the term 'mucormycosis' for the infection produced by Mucorales and introduced the term 'Entomophthoromycosis' to describe subcutaneous phycomycosis caused by Entomophthorales fungi.

Chick et al. (1958) first described amphotericin B as a therapeutic agent for human zygomycosis experimentally produced in rats with the inoculation of *Rhizopus arrhizus* (*R. oryzae*).

In India, *Rhizopus* was first detected as an etiological agent of bovine mastitis (Monga and Kalra 1971). Other earlier reports on detection of Phycomycetes (*Rhizopus* or *Absidia*) in pigs are available which is based on histopathology (Sadana and Kalra 1973; Chauhan and Sadana 1973). *Entomophthora* was also detected histopathologically in the tissues of a mare suffering from fatal cutaneous mycosis (Chauhan et al. 1972).

4.9.1 Morphology

The branching, wide (10–50 μm) and sparsely septate or coenocytic hyphae are produced by *Rhizopus*. The branching usually takes place at right angle which may vary from 45° to 90°. The root-like structure (rhizoid) and the unbranched sporangiophore in clusters (umbels) arise from the opposite side of the hyphae. The horizontal aerial hypha connecting two adjacent umbels is known as stolon. The stolon grows laterally and contacts the growth medium in vitro to expand the colony, whereas the sporangiophore is produced erect to the mycelia. The sporangiophores are dry and easily blown with the wind. They expand at their tip to generate columnella (*columella*) which bears the *sporangium*, containing numerous asexual sporangiospores. The broad area below the columnella is known as *apophysis*, which is inconspicuous (not noticeable) in *Rhizopus* (Fig. 4.19).

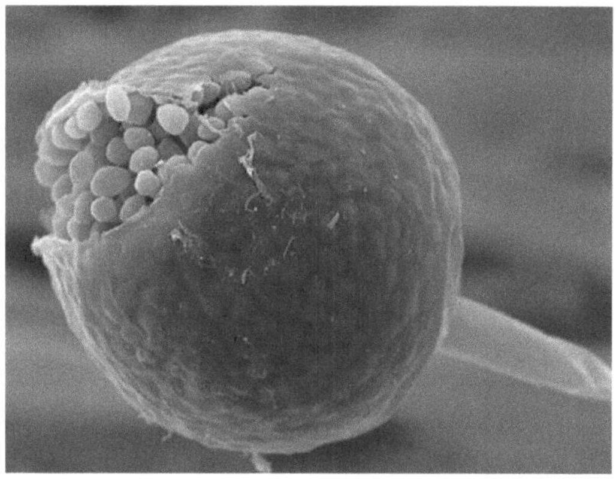

Fig. 4.19 Sporangiophore of *Rhizopus* (Photograph courtesy: Andrii Gryganskyi, Lambert Spawn)

Differentiation of genera is possible on the basis of presence or absence and location of rhizoid, shape of the columnella, size and shape of sporangia and appearance of apophysis. In *Rhizopus*, the rhizoids are found directly opposite the base of aerial sporangiophore. In *Absidia* the rhizoids are also present in opposite but between two adjacent sporangiophores. In *Mucor*, the rhizoids are absent. Further, the stolons are found in *Rhizopus* only. The globose sporangium (round) is produced by both *Rhizopus* and *Mucor*, whereas *Absidia* produces pyriform (pear shaped) sporangium. The apophysis is inconspicuous (not noticeable) in *Rhizopus*, absent in *Mucor* and prominent in *Absidia* (Fig. 4.20).

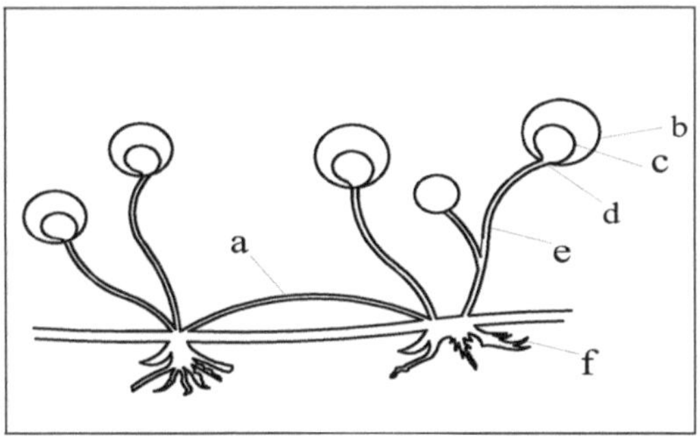

Fig. 4.20 Morphology of *Rhizopus* (schematic); *a* stolon, *b* sporangium, *c* columella, *d* apophysis, *e* sporangiophore, *f* rhizoid

4.9.2 Classification

Before the introduction of fungi kingdom, all the agents causing zygomycosis and entomophthoromycosis were classified under the class Phycomycetes of the subdivision Thallophyta in the plant kingdom. Later, the phylum Zygomycota was introduced under the kingdom Fungi which consists of class Zygomycetes. The Zygomycetes class is composed of two orders, i.e. Mucorales (currently elevated to the subphylum Mucoromycotina) and Entomophthorales (entemon = insect). The order Mucorales contains 8 genera such as *Rhizopus*, *Absidia*, *Mucor*, *Rhizomucor*, *Apophysomyces*, *Cunninghamella*, *Saksenaea* and *Cokeromyces*, whereas the order Entomophthorales contains two genera, i.e. *Basidiobolus* and *Conidiobolus*. Molecular phylogenetic analysis also confirmed the phylum Zygomycota to be polyphyletic. Certain taxa conventionally classified in Zygomycota are now placed under the new phylum Glomeromycota and subphyla incertae sedis (uncertain placement).

Major species under the genus *Rhizopus* are *R. arrhizus* (*R. oryzae*), *R. rhizopodiformis*, *R. azygosporus*, *R. stolonifer*, *R. schipperae*, *R. microsporus* var. *microsporus*, *R. microsporus* var. *rhizopodiformis* and *R. microsporus* var. *oligosporus*. Further, *R. oryzae* can be subdivided into two cryptic species groups, designated as *R. oryzae s. s.* and *R. delemar*.

4.9.3 Reproduction

Most of the *Rhizopus* is heterothallic except *R. sexualis* and *R. homothallicus* which are homothallic in nature. Both heterothallic and homothallic *Rhizopus* can undergo sexual reproduction to produce zygospores. Two sexually compatible hyphae (zygophore) attach with each other by conjugation. The zygophores enlarge to produce progametangium which matures into gametangium. The gametangia are attached with zygophores by a suspensor. There is cytoplasmic continuity between the suspensor and gametangia through the pores (plasmodesmata) for supply of the nutrients. Soon after conjugation, plasmogamy, karyogamy [between (+) and (−) mating types] and meiosis take place which generates zygospores (meiospores). In *Rhizopus*, some nuclei fuse, while others degenerate and meiosis is delayed until zygospore germination ('*Rhizopus* pattern' of zygospore production).

However, *germination* of zygospores for the development of new progeny is a rare event in *Rhizopus* (except in *R. stolonifer*, *R. microsporus*). In *R. stolonifer*, two types of zygopsore germination occur. If the immatured zygospores germinate, they produce a vegetative mycelium which is either (+) or (−). When the matured zygospores germinate, they form a germosporangium that produces spores with equal proportions of (+) and (−) mating types.

The sexual mating is regulated by divergent alleles of a single gene, i.e. *SexM* (−) and *SexP* (+). The *sex* locus is a member of the high mobility gene (HMG) family and is located between two flanking genes that code for a triose phosphate transporter homolog (TPT) and an RNA helicase. In *R. oryzae*, the (+) allele of *sex* locus is significantly larger than (−) allele due to additional gene insertion. However, like the pathogenic function of *MAT* locus in *Cryptococcus*, the role of *sex* locus in *Rhizopus* virulence is not confirmed.

4.9.4 Susceptibility to Disinfectants

Rhizopus is detected to be susceptible to benzalkonium chloride.

4.9.5 Natural Habitat and Distribution

The Mucorales are ubiquitous in nature, thermotolerant and are usually found in decaying organic matter, soil (cultivated grassland, forest and volcanic mud), dung, sewage, municipal waste, saw dust, wood pulp, different foods stuffs, cereals and vegetables such as old bread ('bread mould'), barley, sorghum, wheat, corn, oat, rice, onions, cotton, groundnuts, sweet potatoes, pecans, brazil nuts and tomatoes especially which are rich in sugar and nitrogen. The skin scrapings of healthy fowl are found to harbour *R. microsporus*. Some nonpathogenic strains of *R. oryzae* are used as fermented starters or alcoholic beverages in Asian countries.

Mucormycosis is worldwide in distribution in both developed and developing countries. *R. oryzae* has been identified in India, Pakistan, New Guinea, Taiwan, Central and South America, Africa, Iraq, Somalia, Egypt, Libya, Israel, Turkey, Spain, Italy, Hungary, Czechoslovakia, Germany, Ukraine and the United States, whereas *R. stolonifer* is detected in tropical and subtropical environments.

4.9.6 Genome

The genome of *R. oryzae* is 45.3 Mbp in size and 20 % of the genome is composed of transposable elements (repetitive DNA sequences). There are approximately 13,895 protein-encoding genes which do not overlap with transposable elements. The phylogenetic analysis revealed ancestral whole genome duplication which increases the expression of genes related with virulence, fungi-specific cell wall synthesis and signal transduction. This genome duplication also allows the fungi to grow under various adverse conditions. The genome contains two types of lactate dehydrogenase (LDH) encoding genes, i.e. *ldhA* and *ldhB*. Type I strains of *R. oryzae* produce lactic acid only and contained both *ldhA* and *ldhB*, whereas type II strains produce fumaric and malic acids and possess *ldhB* gene only.

The mitochondrial genome size of *R. stolonifer* is 54.2 kbp.

4.9.7 Isolation, Growth and Colony Characteristics

Rhizopus prefers to grow in standard fungal and bacterial isolation medium such as Sabouraud dextrose agar, malt agar, potato agar, blood agar and chocolate agar at 25–30 °C. It is able to grow well at a wide temperature (up to 40 °C) and pH range (4–9). The virulent strains are able to grow at 37 °C. The media should contain antibiotics to prevent the growth of contaminants except cycloheximide. Most of the Zygomycetes are sensitive to cycloheximide. Isolation of *Rhizopus* from tissue specimens is difficult if the isolation process involves tissue maceration which causes destruction of fungi. As there are no septa, the cytoplasm is expelled out under pressure which prevents the fungal isolation in the media.

Rhizopus produces fluffy white, grey-coloured colonies in the beginning which later becomes yellowish or brownish colonies dotted with black sporangia. The colonies rapidly fill the plates within 1–7 days.

4.9.8 Antigenic Characteristics

The heat-stable, extracellular polysaccharide (EPS) antigen of *Rhizopus* has an immunodominant carbohydrate epitope constituted with 2-*O*-methyl-D-mannose residue. This is a common compound detected in *Rhizopus*, *Mucor*, *Absidia* and *Rhizomucor* which is responsible for the cross-reactivity between these genera.

Rhizopus also secretes rhizoferrin, a siderophore that belongs to the polycarboxylate family. This siderophore can elicit immune response.

Table 4.20 Virulence mechanisms and factors possessed by *Rhizopus*

Virulence factors	Functions
Thermotolerance	It allows *Rhizopus* to grow at host body temperature
Ketone reductase	It helps the fungi to grow in acidic pH and glucose-rich condition such as diabetes mellitus/grain overload
Protease, lipase	They help in tissue invasion by the fungi
Iron acquisition system: The system allows to uptake the iron from the host as in most of the cases, the iron is bound with carrier protein (transferrin) and become unavailable to the growing fungi. So *Rhizopus* grows poorly in the presence of serum which contains these carrier proteins	
High-affinity iron permease (encoded by *FTR1* gene)	It helps in iron uptake from the host which is required for survival of the fungi. It is a part of a reductive system containing redundant surface reductases that reduce ferric into the more soluble ferrous form. The reduced ferrous iron is captured by a protein complex consisting of a multicopper oxidase and a ferrous permease
Rhizoferrin	It is a siderophore which also provides iron to *Rhizopus* through a receptor-mediated, energy-dependent process. It can chelate the ferric iron. *Rhizopus* can also utilise siderophores secreted by other organisms (xenosiderophores) such as deferoxamine (a bacterial siderophore)
Haeme oxygenase	It helps in obtaining iron from host haemoglobin and also promotes the angioinvasion by *Rhizopus*
SreA	It is a transcriptional regulator which is required for adaptation to the ambient iron availability
Toxin	
Rhizoxin (*R. microsporus*, *R. chinensis*)	It is an antimitotic macrocyclic polyketide metabolite and is involved in pathogenesis in plants. Recent studies indicated that instead of *Rhizopus*, this toxin is synthesised by intracellular symbiotic bacteria (*Burkholderia*)
Agroclavin (*R. arrhizus*)	Toxic for ruminants
Rhizonin A (*R. microsporus*)	The toxin causes hepatitis in ducks and rats. The toxin is produced by *Burkholderia*, associated with the fungi *Rhizopus microsporus*

4.9.9 Virulence Factors

The virulence factors of *Rhizopus* are enlisted in Table 4.20.

4.9.10 Transmission

The transmission of *Rhizopus* occurs by inhalation, ingestion and percutanoeus route. The sporangiospores, produced by *Rhizopus*, are easily aerosolised due to small size (3–10 μm in diameter) and are dispersed throughout the environment. They can enter through inhalation, damaged skin due to trauma, burn or perforation produced by venous catheters. Even there is a report of human infection which occurred from the spore-contaminated adhesive bandages. Ingestion route of transmission by mouldy feedstuffs is observed in animals with trauma in buccal cavity. Sporadic cases of transmission by insect bite, intramuscular injection are also reported in human.

The predisposing factors for establishment of infection include malignant haematological disorder (especially acute myelogenous leukaemia) with or without stem cell transplantation, neutropenia ($<0.10 \times 10^9$/L in human), diabetes mellitus (type 1, type 2 and secondary) or grain overload in animals, iron overload (deferoxamine therapy in human), trauma, use of corticosteroids (producing defect in macrophage or neutrophil function) and intravenous drug, neonatal prematurity and malnourishment. The iron overload helps in growth of the fungi. *Rhizopus* can utilise deferoxamine as siderophore for iron uptake. The prolonged use of certain antifungal (voriconazole) in haematological disorder patients receiving stem cell transplants is also associated with mucormycosis. In India, majority of human mucormycosis cases (74 %) are associated with prolonged uncontrolled diabetes mellitus, and a proportion of cases are detected in renal transplant recipients.

The seasonal variation in *Rhizopus* infection in human is also detected. The studies in Asian countries (Israel, Japan) identified the autumn season (August–September) as most suitable for transmission of Mucorales infection.

4.9.11 Pathogenesis

The entry of sporangiospores into the immunocompetent host cannot ensure the establishment of infection. If the host is immunosuppressed due to prolonged use of steroids or the host is suffering from diabetes and satisfy the other predisposing factors, there is every possibility of establishment and progress of infection. Under favourable conditions such as low pH, increased iron content and high glucose concentration, the sporangiospores germinate and produce hyphae. The low pH (acidosis) impairs the iron-binding capacity of serum transferrin and increases the level of free iron which favours the fungal growth. So, diabetic ketoacidosis (DKA) patients suffer more from mucormycosis.

The hyphae can penetrate the blood vessels especially the arteries (angioinvasion), causing thrombosis and necrosis. The endothelial glucose-regulated protein (GRP78/BiP/HSPA5) acts as receptor for *R. oryzae*, although the fungal ligand is still not identified. The interaction between the ligand and receptor helps in endocytosis of hyphae. The endocytosis of both live and killed hyphae produces endothelial cell damage, indicating the presence of a toxic chemical in the hyphae which is not identified. As a consequence, the vascular smooth muscle cells are exposed releasing tissue factors and producing intravascular thrombosis.

The proliferation and dissemination of the fungal hyphae depends on production of virulence factors such as ketone reductase, proteases, lipases and iron acquisition system (Table 4.20). The angioinvasion helps in haematogenous dissemination of the fungi into the different target organs such as the uterus of the immunosuppressed dam causing mycotic abortion.

4.9.12 Disease Produced

The major animal and human diseases produced by *Rhizopus* are enlisted in Table 4.21.

Table 4.21 Major diseases of animal and human caused by *Rhizopus*

Fungi	Host	Disease
Rhizopus	Ruminants	Mycotic abortion and placentitis
	Pigs	Gastric ulcer, submandibular granuloma, abscess in the stomach, lymph node, liver
	Dogs and cats	A rare infection in dogs and cats and is associated with use of steroids, intestinal trauma, diabetes, malnutrition and feline panleukopenia infection. The fungal menace causes vomition, diarrhoea, necrotic lesions in the intestinal wall, peritonitis, intestinal obstruction
	Chicken, duck	The vital organs such as the lungs, heart, aorta, spleen, liver, air sacs, kidney and vertebrae are involved. White nodules are observed in the affected tissues
Rhizopus stolonifer (with *Aspergillus niger*)	Horse	Equine invasive pulmonary mycosis
	Human	In human, mucormycosis produces six major clinical forms such as rhinocerebral, pulmonary, cutaneous, gastrointestinal, disseminated and rare forms, like endocarditis, osteomyelitis, peritonitis and renal infection. The most common sites of invasive mucormycosis are the sinuses (39 %), lungs (24 %) and skin (19 %). The skin and gastrointestinal tract are also commonly infected in children. The mortality rate is 66 % in patients with malignancy, 44 % in patients with diabetes and 35 % in the patients with no other menace. In India, rhino-orbito-cerebral manifestations are most common feature followed by cutaneous disease, and most of the infections are associated with diabetes mellitus

4.9.13 Immunity

The ciliated bronchial cells and mucus in the respiratory tract are the first line of defence. The ciliary movement and the cough attempt to expel the sporangiospores of *Rhizopus* during inhalation. The pulmonary alveolar macrophages act as second line of defence which can phagocytose the sporangiospores before their germination. The intact skin and mucous membrane also act as barrier to prevent the entry of the spores.

The toll-like receptors (TLR 2) and other pattern recognition receptors (PRR) present in the phagocytes can interact with the fungal patterns (PAMP) in the spores or hyphae. The interaction can transmit the intracytoplasmic signal for pro-inflammatory cytokine release and activation of antifungal activities of the phagocytes. The phagocytes can damage the spores or hyphae through oxygen-independent or oxygen-dependent pathways. The study indicated that in comparison to *Aspergillus fumigatus*, the *R. oryzae* spores are less killed by healthy human neutrophils, whereas the magnitude of cytokine release (IL6, IL8, TNFα) from blood monocytes is more in *R. oryzae*-induced cells than *A. fumigatus*. The probable reason is that the cell wall of *R. oryzae* contains more chitin than other fungi which is considered as a potent PAMP for the phagocytes. Further, the antifungals such as echinocandins and amphotericin B formulations exert potent immunomodulatory effect on the phagocytes against *Rhizopus*.

4.9.14 Diagnosis

4.9.14.1 Clinical Specimen

The clinical specimens from animals include vaginal discharge, aborted foetal content, faeces and the vital organs collected after postmortem, whereas the specimens from human include nasal aspirates, sputum, bronchoalveolar lavage, pleural fluid, transbronchial biopsy, blood and skin scrapings.

4.9.14.2 Laboratory Examination

1. *Direct examination*: The direct examination in clinical specimens for detection of *Rhizopus* is vital because the Zygomycetes spores are common in the environment and may produce contamination in the artificial culture. The typical wide, coenocytic hyphae, branching at right angle, are seen when potassium hydroxide with Parker ink or calcofluor white is added to the material.
2. *Isolation and identification*: Media and incubation condition as described earlier will serve the purpose.
3. *Histopathology*: The haematoxylin and eosin (H&E) or Grocott stains are used for the demonstration of *Rhizopus* in the tissue section. The Grocott stain produces better resolution due to high contrast with minimal background impregnation. The tissue alterations include neutrophilic infiltrate, necrosis, thrombosis, septic infarction and angioinvasion.
4. *Molecular biology*: A molecular diagnostic technique based on real-time PCR was developed for the simultaneous detection of *Rhizopus oryzae*, *Rhizopus microsporus* and *Mucor* in both culture and clinical samples. Further, the amplification of the internal-transcribed spacer (ITS2) region by a semi-nested PCR and followed by the hybridisation of the amplicons to the probe using Luminex technology was also performed for simultaneous detection of major medically important fungi including *Rhizopus*.
5. *Matrix-assisted laser desorption/ionisation time-of-flight mass spectrometry (MALDI-TOF-MS)*: Recent development in diagnostic technique includes MALDI-TOF-MS-based system which shows higher accuracy than microscopic methods at reduced time. The differentiation of filamentous fungal species including *Rhizopus* has been established with this cutting edge technique.

4.9.15 Treatment

In human, treatment with amphotericin B lipid formulations is the drug of choice against Zygomycosis. If the treatment does not produce satisfactory result, the combined antifungal therapy consisting liposomal amphotericin B with either caspofungin or posaconazole is preferred.

The antifungal voriconazole is contraindicated in the treatment of Zygomycosis. Sometimes, the surgical intervention (lobectomy or pneumonectomy) and immune reconstitution with colony-stimulating factor (CSF) to increase the blood neutrophil level is required for the survival of the patients.

In animals, systemic antifungal therapy with amphotericin B is preferred to cure the *Rhizopus* infection, although the prognosis is grave even after early diagnosis of the infection. Decreased exposure to the sporangiospores can prevent the infection in animals with adequate ventilation in the barns and reduced feeding of moldy hay or silage.

4.10 *Mucor*

The first description of *Mucor* was observed in Robert Hooke's *Micrographia* (1665) and in Marcello Malpighi's *Anatome Plantarum* (1679). Louis Pasteur noted the fermentative capacity of *Mucor* and described the dimorphism of *Mucor racemosus*. In 1918, Ernst first reported *Mucor corymbifer* (*Absidia corymbifer*) as an etiological agent of human vocal cord infection and established the pathogenic status of *Mucor*.

In India, Barua and Ahmed (1963) first detected *Mucor* from the stomach content of aborted foetus in cattle as an etiological factor of mycotic abortion. *Mucor* was also isolated from the cases of bovine gastritis affecting all the compartments of stomach (Damodaran et al. 1976).

4.10.1 Morphology

The hyphae are wide, coenocytic, branched and usually tapering. The branching occurs at frequent but unpredictable interval along the hyphal length. The sporangiophores are hyaline and rhizoids are usually absent (Fig. 4.21).

Fig. 4.21 Sporangiophore of *Mucor* (schematic)

The property of dimorphism separates *Mucor* from other members of Zygomycetes. The dimorphism is produced by some species of *Mucor* such as *M. racemosus*, *M. rouxii*, *M. genevensis*, *M. bacilliformis* and *M. subtilissimus* (certain strains). The yeast cells can develop from sporangiospores, arthrospores and hyphae but not directly from the zygospores. The *Mucor* yeasts are large (20 μm in diameter), spherical, multinucleate cells, and they produce several buds at random locations. The yeast cell wall is much thicker, more diffuse and more fibrous than the cell wall of hyphae. The yeast cell wall contains two separate layers, whereas the hyphal cell wall is uniform and single layered. There are several environmental factors influencing dimorphism such as anaerobic condition which favours the generation of yeast cells.

Mucor produces both asexual and sexual spores such as arthrospores, sporangiospores and zygospores for different purposes. The arthrospores are produced under adverse climatic and nutritional condition after cessation of logarithmic growth. The arthrospores are produced through septation of coenocytic hyphae and the deposition of a new three-layered wall beneath the original hyphal wall. The arthrospores produced by dimorphic *Mucor* are able to germinate into hyphae or yeast cells. The zygospores are spiny, thick-walled, black-coloured structure produced by the hyphal fusion of either heterothallic or homothallic mating types.

The *Mucor* sporangiophores expand at its tip to generate columnella (*columella*) which bears the sporangium (15–80 μm in diameter), containing numerous asexual sporangiospores (like *Rhizopus*). The *apophysis* is absent in *Mucor*. After maturity, the sporangial wall deliquesces, and the spores remain attached with the columella, so that they could not be carried away with the wind. The sporangiospores are ellipsoidal, uninucleate (except *M. mucedo*) in nature, but they vary in size according to the species. *M. bacilliformis* produces the smallest and *M. mucedo* produces the largest sporangiospores. The cell wall of the sporangiospores is enriched with protein, lipid, melanin and glucan. The melanin and glucan are absent in hyphae, whereas the galactose and fucose present in hyphal cell wall are not detected in sporangiospores. The sporangiospores help in the transmission of infection. They germinate to produce hyphae or yeast cells. The spores first undergo a logarithmic growth phase (spherical growth) due to macromolecular synthesis. During this growth phase, a new vegetative cell wall is synthesised under the spore cell wall. The chemical constituents of the vegetative cell wall are specific for either yeast or hyphae in dimorphic *Mucor*.

4.10.2 Reproduction

Most of the *Mucor* species are able to reproduce sexually by heterothallic (*M. racemosus*, *M. mucedo*, *M. hiemalis*) or homothallic mating (*M. rouxii*, *M. genevensis*). In heterothallic mating, fusion of gametangia occurs between opposite mating types. In homothallic mating, fusion occurs between the gametangia located at the different sites of the same thallus. The zygophores develop at the end of the hyphal branches. In response of trisporic acid secreted by the opposite mating types from prohormone precursors, the zygophores are attracted to each other. After direct contact between two zygophores, they swell and form progametangia. The progametangia fuse to produce the gametangium, which undergoes plasmogamy and karyogamy. The wall becomes thickened with multiple layers and forms the zygospores. However, the generation of zygospores with karyogamy is rare in nature. The zygospores germinate to produce haploid sporangiospores with uniparental genetic makeup.

Like *Rhizopus*, a similar sex locus was reported in the *M. circinelloides* and *M. mucedo* genome. The putative sex locus contains a single high mobility group (HMG) transcription factor gene which is flanked by an RNA helicase and a triose phosphate transporter gene. The HMG proteins are designated as SexP for the (+) and SexM for the (−) mating type, respectively. The transcription of *SexM* is highly stimulated than *SexP* with the pheromone (trisporic acid) in *M. mucedo* cultures kept in vitro. The higher-transcriptional stimulation is correlated with difference in the promoter sequences of SexM and SexP. However, whether the similar kind of higher-transcriptional stimulation of *SexM* and simultaneously repression of (+) mating type occurs under in vivo condition is yet to be elucidated.

The (−) mating types of *M. circinelloides* can produce larger spores and are more virulent than the (+) mating types.

4.10.3 Classification

The genus *Mucor* is placed under the order Mucorales (described in details in Sect. 4.9). The major species under the genus *Mucor* are *M. circinelloides* complex, *M. hiemalis*, *M. racemosus*, *M. ramosissimus* and *M. rouxianus*. The *M. circinelloides* complex consists of three subspecies, i.e. *M. circinelloides* f. *lusitanicus* (*Mcl*), *M. circinelloides* f. *circinelloides* (*Mcc*) and *M. circinelloides* f. *griseocyanus* (*Mcg*).

4.10.4 Susceptibility to Disinfectants

Mucor are susceptible to sodium hypochlorite (2.4 %), phenol, chloroxylenol and isorophyl alcohol.

4.10.5 Natural Habitat and Distribution

Mucor is saprophyte and ubiquitous in nature. They are commonly detected in soil, hay, dog fur, cereals, nut and flour (*M. circinelloides*, *M. hiemalis*), wheat grains (*M. racemosus*), soyabean seeds, stored cabbage (*M. hiemalis*), oranges (*M. circinelloides*) and environmental samples throughout the world.

4.10.6 Genome

The nuclear genome of *Mucor* (*M. miehei*) contains about 100 copies of rRNA gene, each of which is 7.7–12 kbp in size and is arranged in tandem. The genes encoding the 25S, 18S and 5.8S rRNAs are closely linked within the repeated unit which also contains the 5S gene. This 5S gene appears to be transcribed in the opposite direction. In *M. racemosus* genome, the 35S rDNA cistron and the 5S rDNA are linked but are transcribed from opposite strands.

The transposable element is also detected in nuclear genome of *Mucor racemosus* and other *Mucor* species. In eukaryotes, most of them influence gene expression and promote genetic diversity by mediating genetic rearrangements. Such kind of function is yet to be elucidated in *Mucor*.

The mitochondrial genome of *Mucor* contains a circular chromosome. The approximate size is 63.8 kbp in circumference (*M. racemosus*). The average size of *M. piriformis* mitochondrial genome is 33.6 kbp containing the mitochondrial genes such as *cob*, *nad2* and *SrRNA*. Exceptionally, the *M. racemosus* mitochondrial genome is found to exist in the form of two flip-flop isomers with inverted repeat sequences encoding rRNA genes.

4.10.7 Isolation, Growth and Colony Characteristics

Mucor can be isolated in selective or nonselective fungal media. They are rapid grower, and the mycelia can fill and reach up to the lid of the Petri dish within 1–7 days ('lid-lifters'). The colony characteristics of major *Mucor* species are described in Table 4.22.

Table 4.22 Colony characteristics of *Mucor* species

Mucor species	Colony characteristics
M. circinelloides	The growth takes place at 25–37 °C and the colonies are pale grey or yellowish. The colonies become brownish at 37 °C
M. hiemalis	Optimum growth temperature is 32 °C. Greyish colonies are produced and reverse pigmentation is pale
M. rouxianus	Optimum growth temperature is 37 °C and the required temperature range is 9–45 °C. The colours of the colonies are white, yellow or grey. The colonies are 4 mm in height and delicate in nature
M. racemosus	The colonies are low to moderately raised and are brownish in colour due to formation of sporangia
M. ramosissimus	Rapid growth occurs at 25 °C. The colonies are grey to buff in colour

4.10.8 Antigenic Characteristics

Mucor (*M. racemosus*) possesses a heat-stable, extracellular polysaccharide (EPS) antigen which is shared with *Rhizopus*, *Absidia* and *Rhizomucor*.

4.10.9 Virulence Factors

The virulence factors of *Mucor* are enlisted in Table 4.23.

4.10.10 Transmission

Like other Mucorales, the transmission of *Mucor* occurs by inhalation, ingestion and percutanoeus route. Inhalation of spores results rhinocerebral and pulmonary form of the infection in human. The percutaneous infection may occur during injection, insect bite and trauma. The nail infection with *Mucor* in human was reported to be originated from handling with the contaminated objects (e.g. orange).

Immunosuppression due to bone marrow transplantation, aplastic anaemia, diabetes, asthma, burns, hepatitis and human

Table 4.23 Virulence mechanisms and factors possessed by *Mucor*

Virulence factors	Functions
Thermotolerance	*Mucor* does not show any thermotolerance. Only *M. rouxianus* grows optimally at 37 °C, whereas *M. ramosissimus* and *M. circinelloides* grow slowly and *M. hiemalis* does not grow at all at this temperature
Spore size	Larger spore size of *M. circinelloides* complex is associated with virulence, and generally (−) mating types produce abundant larger spores than the (+) mating types. The larger spores are more virulent probably due to shorter time requirement for germ tube generation and subsequent hyphal growth
Fungal metabolite	It is produced by *M. hiemalis* and it can suppress cell division in germinating grains
Calcineurin	The calcineurin pathway helps in the dimorphic transition from yeast to hyphae, correlated with virulence
Sialic acid	The surface sialic acids contribute to the negative charge of *Mucor* yeasts and spore cells and protect them from phagocytosis

immunodeficiency virus (HIV) infection may act as predisposing factors for *Mucor* infection.

4.10.11 Pathogenesis

Most of the *Mucor* species are opportunistic fungi and can establish the infection during breakdown of the host immunity. Under the low pH, high glucose concentration and increased iron content, the spores germinate and produce hyphae. The hyphal extension can invade the blood vessels, nerves and fascial planes. Sometimes haematogenous dissemination of the spores occurs to establish the infection in different organs such as udder in cattle. In immunocompromised hosts, *Mucor* produces invasive diseases in solid organs such as pulmonary, rhinocerebral, rhino-orbital and gastrointestinal illness.

In immunocompetent hosts, the infection is restricted within the skin and nails. Erythema, papules, nodules and plaques without superficial necrosis are noted in the skin. In the nails, punctate erosions are observed.

4.10.12 Disease Produced

The major animal and human diseases produced by *Mucor* are enlisted in Table 4.24.

4.10.13 Immunity

The ciliated bronchial cells and mucus in the respiratory tract are the first line of defence. The ciliary movement and the cough attempt to expel the sporangiospores of *Mucor* during inhalation. The pulmonary alveolar macrophages act as second line of defence which can phagocytose the sporangiospores before their germination. The spores of *Mucor ramosissimus*, *M. plumbeus* and *M. circinelloides* are also observed to activate alternative complement pathway. The intact skin and mucous membrane act as barrier to prevent the entry of the spores.

4.10.14 Diagnosis

4.10.14.1 Clinical Specimens

The clinical specimens from animals include mastitic milk, skin scrapings and swab from thoracic cavity, etc., depending upon the clinical symptom, whereas the specimens from human include nasal aspirates, sputum, bronchoalveolar lavage, pleural fluid, blood, skin scrapings and nail specimens.

4.10.14.2 Laboratory Examination

1. *Direct examination*: The typical wide, coenocytic hyphae, branching at right angle, are seen when potassium hydroxide with Parker ink or calcofluor white is added to the material.
2. *Isolation and identification*: Media and incubation condition as described earlier will serve the purpose.

Table 4.24 Major diseases of animal and human caused by *Mucor*

Fungi	Host	Disease
Mucor spp.	Human	Pulmonary, rhinocerebral, rhino-orbital, gastrointestinal cases predominate among immunocompromised patients. Gastrointestinal diseases include abdominal pain with hepatic abscesses, necrotic ulcerations of the stomach wall, diarrhoea and paralytic ileus. Skin lesions and nail infections are observed in immunocompetent patients
	Cattle	Mastitis, thromboembolic encephalomyelitis (rare)
	Cat	Non-painful, subcutaneous swelling at nasal dorsum associated with scratch injury
	Duck	Thoracic cavity infection
	Pig	Papular dermatitis (rare)
	Marine animals	(a) *Mucor* is associated with fungal granuloma in cardiac muscles of killer whale coinfected with Herpesvirus and *Aspergillus* (b) *M. circinelloides* is detected to produce cardiac tissue damage in juvenile zebra shark (c) Systemic mucormycosis was detected in a captive hooded seal comprising muscle necrosis over the left flank, necrotic iliac lymph node and necrotic bronchial lymph node
	Toad	Dermatitis
M. ramosissimus	Canary birds	Dermatitis and feather loss
M. amphibiorum	Platypus	Cutaneous lesions ranging from raised red nodules or plaques with purulent material to ulcerated lesions with central cavitation, red exuding centres and raised epidermal margins are observed

3. *Histopathology*: The haematoxylin and eosin (H&E) stain is used for demonstration of *Mucor* in the tissue section. In subcutaneous lesions, coenocytic hyphae are immersed in inflammatory exudate of neutrophils, eosinophils, epitheloid and multinucleated giant cells. In dermis, spores and coenocytic hyphae are observed. The superficial and deep nodular granulomatous dermatitis consisting of lymphocytes, histiocytes, plasma cells and giant cells is detected in epidermis.
4. *Molecular biology*: There are no serological tests available for detection of *Mucor*, so PCR-based molecular biology techniques are better options for detection in clinical specimens. The broad range real-time PCR, DNA microarray is developed for simultaneous detection of several pathogenic fungi such as *Mucor*, *Aspergillus*, *Candida*, *Cryptococcus*, dermatophytes, etc. For rapid detection of circulating DNA from serum of patients, quantitative PCR (qPCR)-based approach was developed. Subsequently, PCR followed by high-resolution melt analysis (PCR/HRMA) is used for the detection of *Mucor* in bronchoalveolar lavage samples.

4.10.15 Treatment

Amphotericin B is the drug of choice for treatment of *Mucor* infection in human and animals. *M. ramosissimus* is detected to be susceptible to miconazole in vitro. The subcutaneous lesions and the nail infection can be treated with potassium iodide and mercuric chloride (1: 10,000).

4.11 *Penicillium*

Link (1809) introduced the name *Penicillium* which was derived from the word ‘penicillus’ (little brush).

Penicillium marneffei was first isolated from a hepatic lesion of bamboo rats (*Rhizomys sinensis*) in 1956 which were maintained in captivity for experimental infections at Pasteur Institute of Indochina, South Vietnam (Capponi et al. 1956). The fungus was named *Penicillium marneffei* in the honour of Hubert Marneffe, the then director of Pasteur Institute of Indochina. The first reported human case of *P. marneffei* was accidental when the scientist Dr. G. Segretain (France) pricked his own finger with a needle filled with the culture. He developed small nodule at the site of inoculation which was followed by axillary lymphadenopathy. The infection was treated successfully with the antifungals (Segretain 1959). In 1973, first naturally occurring human infection was reported which occurred in an American patient, suffering from Hodgkin's disease and living in Southeast Asia (DiSalvo 1973). Its potentiality as an opportunistic pathogen was observed during human immunodeficiency virus (HIV) pandemic (1988 onwards) in Southeast Asian countries. The maximum number of *P. marneffei* infection in human with AIDS was diagnosed in Thailand along with other countries such as Cambodia, China, Hongkong and India.

Sir Alexander Fleming (1929) discovered the first antibiotic penicillin during accidental contamination of bacterial cultures with *Penicillium notatum* at Saint Mary's Hospital, United Kingdom.

4.11.1 Morphology

The mycelium of *Penicillium* consists of highly branched, hyaline and septate hyphae (Fig. 4.22). Sometimes the hyphae interweave with each other to form rope-like structure. The conidiophores are branched. The secondary branches of conidiophores are known as *metulae* which bear the *phialide* from which the smooth or rough and round conidia (2.5–5 μm) are borne (Fig. 4.23).

Peroxisomes are morphologically simple organelles present in *Penicillium* (e.g. *P. chrysogenum*) and other filamentous fungi which produce secondary metabolites such as penicillins, polyketides and terpenes. The organelles are produced from endoplasmic reticulum and are detected predominantly at the hyphal tips. The formation of peroxisomes is controlled by peroxins, encoded by *PEX* gene.

P. marneffei is a dimorphic species under the genus *Penicillium* showing filamentous growth and asexual reproduction (conidia) at 25 °C and generation of yeast cells at 37 °C. The yeast cells are characterised by the formation of

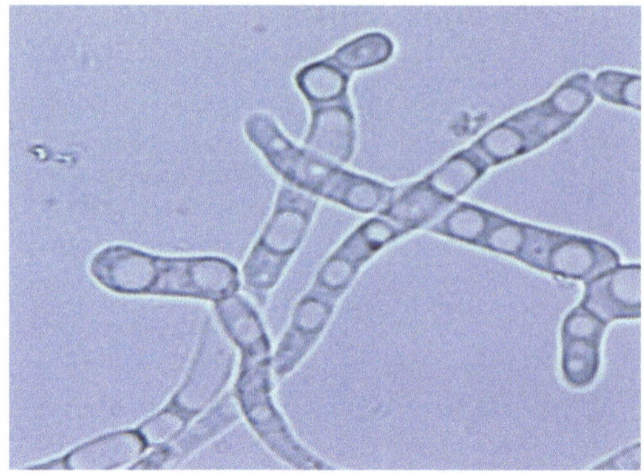

Fig. 4.22 Hyphae of *Penicillium chrysogenum* (Photograph courtesy: Marco van den Berg, DSM Biotechnology Centre, The Netherlands)

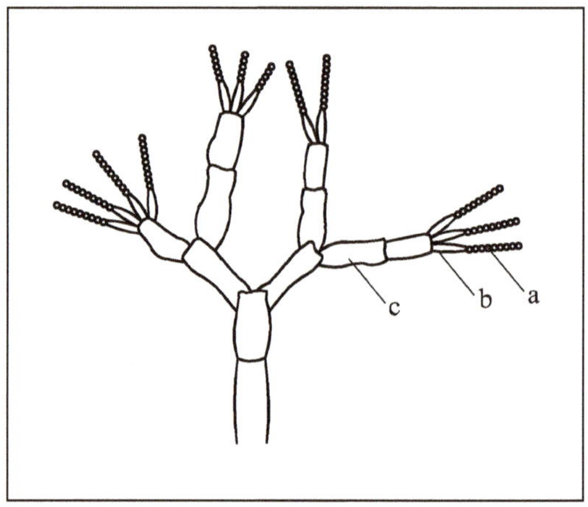

Fig. 4.23 Conidial arrangement of *Penicillium* (schematic); *a* conidia, *b* phialide, *c* metulae

intracellular fission cells within macrophages which represent the parasitic phase.

4.11.2 Life Cycle

The conidium of *P. marneffei* germinates (25 °C) to produce germ tube. The germ tube grows apically to make hypha. The subapical cells grow at a new point in the hypha to make its branches. After formation of the mycelia, the specialised vegetative hyphal cells (*foot cell*) differentiate to produce multinucleate aerial stalk cells. The tips of these stalk cells grow by budding to generate uninucleate metulae and phialides. The first uninucleate bud is a metula which subsequently buds to produce more sporogenous cell type, known as phialide. These phialides produce conidia in basipetal mode, i.e. old spores are replaced with new spores. Thus, a chain of conidia is produced having green colouration. Under carbon-limited condition, short unbranched stalks with conidia are produced or sometimes the metulae appear directly from the foot cells.

The conidia can germinate and produce highly branched hyphae at 37 °C. These hyphal cells are shorter (20 μm) than vegetative hyphal cells (40 μm) produced at 25 °C and are separated by double septa. This is a transitory phase and these hyphal cells are known as 'prearthroconidial cells'. These prearthroconidial cells generate arthroconidia (uninucleate) by dissolving the double septa within 48 h. The arthroconidia divide by fission to generate true yeast cells within 72 h. This kind of arthroconidiation followed by yeast formation is unique in dimorphic *P. marneffei*. Within the body of the host, the yeast cells may remain as intracellular (within macrophages) or extracellular. The intracellular yeast cells are oval or spherical and smaller (2–3 μm in diameter). The extracellular yeast cells are elongated and larger (13 μm in diameter).

The yeast cells can undergo coupled nuclear and cytoplasmic divisions. When the yeast cells are shifted at 25 °C, the nuclear division becomes uncoupled. The elongated multinucleate cells divide only by septation without cell separation, and they produce branched hyphal network (yeast to hyphae transition).

The dimorphic switch in *P. marneffei* is regulated by temperature which causes cellular changes such as coupling and uncoupling of cellular and nuclear divisions, septation with or without cell separation, etc.

4.11.3 Classification

Penicillium belonged to Trichocomaceae family under the order Eurotiales (class Eurotiomycetes and phylum Ascomycota). The *Penicillium* is a polyphyletic genus and the species can be divided into two major clades, i.e. *Penicillium sensu stricto* (previously *Penicillium* subgenus of Pitt's classification) and *Talaromyces* (previously *Talaromyces* and *Penicillium* species under the subgenus *Biverticillium*). *Penicillium sensu stricto* currently includes the genera *Eupenicillium*, *Chromocleista*, *Eladia*, *Hemicarpenteles*, *Thysanophora* and *Torulomyces*. Some authors prefer to treat this clade under the singular genus *Penicillium*. There are total 225 accepted species under the genus *Penicillium*. Major important species are *Penicillium marneffei*, *P. chrysogenum*, *P. commune*, *P. cyclopium*, *P. lilacinum* and *P. oxalicum*.

Penicillium is pleomorphic fungi detected either separately as anamorph (asexual), teleomorph (sexual) or the holomorph (both phases). A few species have only one known state. However, some authors segregated sexually reproducing *Penicillium* (teleomorph) into the genus *Talaromyces*, *Hamigera* and *Eupenicillium*.

Modern taxonomy is based on fungal genome sequencing and genealogical concordance phylogenetic species recognition (GCPSR)-based analysis of sequence data. It changed nomenclature of certain important *Penicillium* species. For example, the major pathogenic species (*Penicillium marneffei*) is currently designated as *Talaromyces marneffei*. The original strain of Fleming was later identified as *Penicillium chrysogenum*. Recent taxonomy showed that *Penicillium chrysogenum* is a complex of five species, i.e. *P. chrysogenum*, *P. rubens*, *P. vanluykii*, *P. tardochrysogenum*, *P. allii-sativi*. Fleming's strain and the classical strain used for the penicillin production (Wisconsin strain) are actually *P. rubens*.

There are several genera having *Penicillium*-like conidiophores including *Hamigera*, *Paecilomyces*, *Rasamsonia*, *Sagenomella*, *Talaromyces* and *Trichocoma* which are also known as *Penicillium*-like anamorphs.

4.11.4 Reproduction

Penicillium produces asexual spores (conidia) by a process known as 'conidiation'. There are four

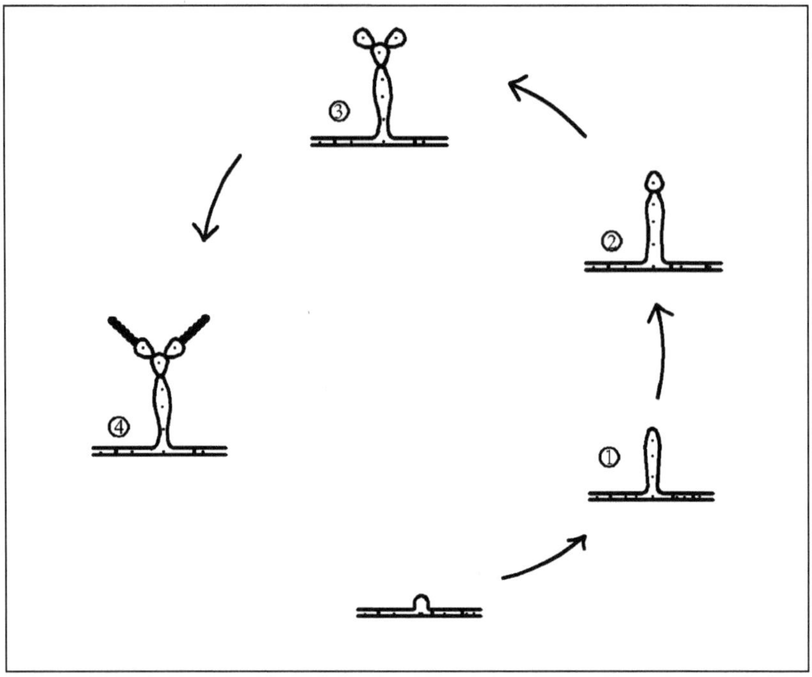

Fig. 4.24 Conidiation stages of *Penicillium* (schematic)

stages of conidiation. In the first stage, the extension of apically growing hyphae is seized which is followed by the formation of septum to separate the apical cell. The apical cell is swelled and there is subapical branching (stage 2). The apical cell differentiates into phialide (stage 3) which buds at the tip to generate the conidium (stage 4) (Fig. 4.24). Upon the first conidium, new conidia appear at hourly interval to produce a chain of conidia. The whole procedure up to stage 4 takes approximately 7 h time. From the subapical branches, formation of phialides and conidia takes place to generate the typical brush-like structure ('penicillus').

Certain environmental factors regulate conidiation in *Penicillium* such as exposure to air, light, high osmolarity, calcium and nutrient starvation. The hydrophobin coating of aerial hyphae prevents the drying during exposure to air. Light is not a major requirement for conidiation as some species of *Penicillium* can produce conidia without light. The calcium can bind the *Penicillium* hyphae extracellularly to trigger the induction of conidia by an unexplored mechanism. Nutrient starvation is detected by several protein sensors in the hyphae which can transmit a positive signal for conidiation. There are certain endogenous inducers such as conidiogenone, conidiogenol, sporogen (PF1) which may induce the conidiation through quorum sensing, a communication network detected in bacteria. The sporogen can act as inducer in darkness also. The conidiation signalling pathway is regulated by GasA G protein α-subunit in *P. marneffei*. The signalling involves a cAMP-dependent protein kinase A cascade.

Majority of *Penicillium* species (73 %) including *P. marneffei* do not have any sexual state. However, the presence of *stlA* gene (homologue of conserved *STE12*) suggests a cryptic sexual cycle in *P. marneffei*. The sexual cycle is apparently not visible probably due to heterothallic nature of *P. marneffei*. Heterothallic fungi require two compatible mating types for pairing which may not be available in the environment simultaneously. Majority of other *Penicillium* species having sexual reproduction are homothallic in nature. The sexual reproduction is noted in *P. chrysogenum* and *P. dipodomyis*, and experimentally sexual cycle is induced in *P. pinophilum*, previously considered as an asexual species.

4.11.5 Natural Habitat and Distribution

Penicillium and *Aspergillus* are considered as the most prevalent fungi in the world including temperate, tropical as well as arctic locations. Many species of penicillia are soil saprophytes except *P. marneffei*. The saprophytic penicillia can be isolated from rotting fruits, plant roots and indoor environments such as archives, compost heaps, damp buildings, etc. They can liberate different enzymes (phosphatase), chelators, organic anions and siderophores (trihydroxymates, coprogen, ferricrocin) to survive in the soil. Association of penicillia with the plant roots stimulate the plant defence system leading to improved seed germination and growth. For pathogenic penicillia (*P. marneffei*), different species of bamboo rats (*Rhizomys sinensis*, *R. pruinosus*, *R. sumatrensis*, reddish-brown subspecies of *Cannomys badius*) are considered as enzootic reservoir. In India (especially in Manipur, a Northeastern state), *Cannomys badius* acts as major reservoir.

The epidemiological study of *P. marneffei* in human identified Southeast Asia as an endemic zone. The infection was reported from Northeastern India, Thailand, China, Taiwan, Hong Kong, Laos, Cambodia, Malaysia, Vietnam and Myanmar. Soil exposure especially in rainy season was the critical risk factor associated with *P. marneffei* infection.

4.11.6 Genome

The genome of *P. marneffei* contains three chromosomes (2.2, 4, 5 Mbp); however, the size of the genome is 17.8–26.2 Mbp. A section of genome (9 %) is occupied by random sequence tags (RST) containing genes for ribosomal protein, tRNA synthetase, translation initiation, elongation factor, metabolism and multidrug resistance. The internally transcribed spacer (ITS) region contains very high GC mol%. The mitochondrial genome (35 kb) has similarity with the mould (*Aspergillus nidulans*) than the yeast.

The *abaA* gene present in *P. marneffei* genome controls the generation of conidia probably through the control of cell cycle during sporulation. The expression of *abaA* gene is increased during conidiation. The *abaA* deletion mutants produce phialide-like cells but are unable to produce conidia. Another asexual reproduction regulatory gene is *stuA* which is a member of the APSES group of transcription factors possessing a basic helix-loop-helix (bHLH) DNA-binding motif. The expression of *stuA* is also increased during conidiation. However, *stuA* deletion mutants are unable to produce metulae and phialides, but the conidia are generated directly from the conidiophores. The other regulatory genes for asexual development and secondary metabolite formation include Gα-subunit gene (*gasA*). There is a *CDC42* homologue (*cflA*) detected in *P. marneffei* which is required for correct morphogenesis of yeast cells and accurate polarisation of hyphae. However, these regulatory genes (*stuA*, *gasA*, *cflA*) have no role in dimorphic switch of *P. marneffei*.

MAT gene was identified in several *Penicillium* species with both MAT1-1 and MAT1-2 genotypes. Other sex-related genes such as *stlA* (homologue of conserved *STE12*), pheromone precursor and receptor genes were detected in *P. marneffei* (within RST) and *P. chrysogenum*, respectively.

Comparative genomic analysis between penicillin-producing strains (*P. chrysogenum*) revealed higher transcription of penicillin biosynthesis genes (*pcbAB*, *pcbC*, *penDE*) in high amount of penicillin-producing strains. Due to high genomic plasticity and ability to grow in different media and conditions, strains of *P. chrysogenum* are the first choice for the production of penicillin enzymes in biotechnology industry.

4.11.7 Isolation, Growth and Colony Characteristics

Penicillium can be isolated on Czapek dox agar, potato dextrose agar and 25 °C. Most species sporulate within 7 days, whereas the mycelial phase of *P. marneffei* can be isolated in Sabouraud glucose agar without cycloheximide at 25 °C. They are converted into the yeast phase at 37 °C in brain–heart infusion agar (thermal dimorphism).

The colonies of *Penicillium* are bluish-green and velvety. The colour and texture of the colonies vary with the species. For example, *P. marneffei* produces diffusible wine red or brownish red pigment in the media.

4.11.8 Antigenic Characteristics

Galactomannan (GM) is the major antigen of *Penicillium* which is a carbohydrate molecule composed of mannose residues with side chains of β (1, 5)-linked galactofuranosyl residues. It is released through the pores at the growing hyphal tips during logarithmic growth phase of the fungi in highest amount which helps in the detection of the antigen for diagnosis. GM is found in other moulds such as *Aspergillus*, *Fusarium*, *Alternaria*, *Histoplasma* and yeasts including *Cryptococcus* which can produce antigenic cross-reaction.

Pen c is a protective antigen and allergen detected in *P. citrinum* and it is the homologue of Asp f commonly found in *Aspergillus*. The Asp f is a 19 KDa protein and it has two T cell epitopes (11mer and 13mer) which offer protection against the fungi.

The yeast phase of *P. marneffei* contains three immunodominant proteins (50 KDa, 54 KDa and 61 KDa) which are specific for the fungi and can be used in serodiagnosis.

4.11.9 Toxins

4.11.9.1 Ochratoxin

The ochratoxin in feed stuffs in temperate countries is mainly produced by *Penicillium verrucosum* and *P. nordicum*. Growth of *P. verrucosum* occurs in cereals in North and Central Europe and Canada. *P. nordicum* prefers to grow in meat and meat products such as pork. The meat contamination takes place either during ripening (meat processing) or from the animals fed with contaminated cereals. The ochratoxin is considered as potent nephrotoxic, carcinogenic (human carcinogen of 2B group), genotoxic and immunotoxic.

4.11.9.2 Citrinin

Citrinin is a fungal metabolite which was first isolated from *Penicillium citrinum*. Other species such as *P. expansum* and *P. verrucosum* can also produce citrinin. It is frequently detected in feed stuffs or food along with ochratoxin A and it can increase the toxicity of ochratoxin synergistically. The intoxication with both of the toxins may cause endemic nephropathy. Further, citrinin is also embryocidal and foetotoxic probably due to the production of oxidative stress or increased permeability of mitochondrial membranes. Oral LD_{50} for rats is 50 mg/kg body weight, while subcutaneous LD_{50} is 67 mg/kg body weight. The heat-decomposed products of citrinin (CTN H_1 and CTN H_2) are more toxic than the parent citrinin.

4.11.9.3 Ribotoxin (RNase T1)

Certain *Penicillium* species can produce a ribotoxin known as RNase T1. Like other ribotoxins, RNase T1 can also cleave the phosphodiester bond located within a universally conserved sequence of the large rRNA gene (sarcin–ricin loop). The cleaving inhibits protein synthesis and causes cellular death. The virus-infected or virus-transformed cells having altered cellular membrane are more susceptible to ribotoxin. The ribotoxins are used in preparation of immunotoxin due to their cytolytic property.

4.11.10 Virulence Factors

The other virulence factors of *Penicillium* are enlisted in Table 4.25.

4.11.11 Transmission

In human and animals, traumatic implantation of the fungi into the skin results infection.

Table 4.25 Virulence mechanisms and factors possessed by *Penicillium marneffei*

Virulence factors	Functions
Catalase-peroxidase protein (encoded by *cpeA*)	The high expression of *cpeA* contribute to the survival of *P. marneffei* within the host cells at 37 °C
Acid phosphatase	The enzyme produced by *P. marneffei* helps in survival of the fungi within host macrophages
Vacuolar serine proteases (cerevisin) such as Pen c2, Pen c18, Pen ch18, Pen o18	Vacuolar serine proteases (cerevisin) are produced by *P. citrinum* (Pen c), *P. chrysogenum* (Pen ch) and *P. oxalicum* (Pen o). They act as fungal allergen by activating other proteinaceous allergens present in the inoculum and make the whole organism a potent allergen (bystander effect)

Additionally *P. marneffei* is also transmitted by inhalation of the contaminated dust.

4.11.12 Pathogenesis

After entry within the body of the host, *P. marneffei* is converted into the yeast cells and spreads by haematogenous route throughout the body. The survival within the phagocytes such as macrophages depends on the production of acid phosphatase and iron availability. The severity of infection depends on immune status of the host. In immunocompromised patients, the yeast cells can multiply and kill the macrophages with dissemination throughout the body and produce fatal necrotising reactions and histiocyte infiltrations. Severe *P. marneffei* infections can affect different tissues and organs such as the bone marrow, liver, spleen, kidney, lungs, lymph nodes, skin and soft tissues, whereas in immunocompetent patients, granulomatous and suppurative reactions are detected in the lung, skin, liver and subcutaneous tissues with the formation of multiple abscesses.

4.11.13 Disease Produced

The major human and animal diseases produced by *Penicillium* are enlisted in Table 4.26.

4.11.14 Immunity

The conidia of *P. marneffei* are attached with the respiratory epithelial cells with the help of a sialic acid-containing receptor before phagocytosis. The phagocytic cells act as primary line of defence against the fungi. The engulfment of the fungi by the macrophages is a cation-independent process, and it largely depends on the glycoprotein receptor containing *N*-acetyl-β-glucosaminyl groups.

The CD4+ T cells producing Th1 type of immune response play major role in protective immunity against *P. marneffei*. The T cells activate the macrophages with the help of IFNγ and other cytokines. The activated macrophages can kill the intracellular yeast cells through L-arginine-dependent nitric oxide pathway. The neutrophils after activation with the granulocyte-macrophage colony-stimulating factor (GM-CSF) can lyse the yeast cells (not conidia) with the secretion of different granular cytolytic molecules.

4.11.15 Diagnosis

4.11.15.1 Clinical Specimen

The human clinical specimens for diagnosis of *P. marneffei* infection include bone marrow aspirate, blood, skin scrapings, sputum, broncho-alveolar lavage, pleural fluid, cerebrospinal fluid, pharyngeal ulcer scrapings, palatal papule scrapings, urine, faeces and lymph node biopsies, skin biopsies, liver biopsies, kidney, pericardium, stomach or intestine.

The animal clinical specimens include mastitic milk, skin scrapings, aborted foetus.

4.11.15.2 Laboratory Examination

1. *Direct examination*: Detection of *Penicillium* is possible by direct examination of the

Table 4.26 Major diseases of animal and human caused by *Penicillium*

Fungi	Host	Disease
Penicillium marneffei	Human	The infection is confined in Southeast and Eastern Asia, including northeast India, Thailand, China, Hong Kong, Laos, Cambodia, Malaysia, Myanmar, Vietnam and Taiwan. The syndrome includes fever, weight loss, cough, anaemia (haemoglobin level of 10 g/dl or less), lymphadenopathy and hepatosplenomegaly. In HIV-infected patients, skin lesions appear having similarity with molluscum contagiosum in most of the cases (60–85 %). The skin lesions appear on the face, upper trunk and extremities. The lesions are necrotic or generalised papules or nodules and with central umbilication. The osteoarticular lesions are sometimes observed in the ribs, long bones, vertebrae and skull. Without the skin lesions, clinically it is difficult to diagnose because lots of opportunistic pathogens produce a similar kind of syndrome. The bone lesions are also detected in tuberculosis, cryptococcosis, blastomycosis and African histoplasmosis
Penicillium sp.	Ruminants and other animals	Mastitis, keratitis, dermatitis, abortion and other reproductive disorders
Penicillium vermiculatus	Cattle	Mycotic abortion (rare in occurrence)
P. decumbens	Human	Invasive mycosis in HIV-infected patients

smears prepared from the clinical specimens (blood, bone marrow aspirate, biopsies) and stained with Wright's stain.

2. *Isolation and identification*: Media and incubation condition as described earlier will serve the purpose. The ideal clinical specimens for culture are bone marrow, blood and skin biopsy.
3. *Histopathology*: The haematoxylin and eosin (H&E), Grocott methenamine silver and periodic acid–Schiff (PAS) stains are used for the demonstration of *P. marneffei* in the tissue section. Within the macrophages or histiocytes, round or oval yeast cells are detected. Sometimes they appear as fission arthroconidia which will produce yeast cells by cross wall formation which is considered as differentiating criteria from other yeasts. Extracellular elongated or sausage-shaped cells are also observed.

   In histological sections, indirect fluorescent antibody test can detect *P. marneffei*. In paraffin-embedded formalin-fixed tissues, *P. marneffei* can be detected with monoclonal antibody against the galactomannan antigen (EBA1). Immunoperoxidase staining can detect the fungi in deparaffinised tissue sections of skin biopsies.
4. *Detection of fungal antigen*: The fungal antigen can be detected with immunodiffusion test, latex agglutination test, monoclonal antibody-based sandwich ELISA.
5. *Serological tests*: The antibodies developed in patients against *P. marneffei* infection can be detected by immunodiffusion, indirect fluorescent antibody test (detects IgG) and ELISA with a recombinant mannoprotein (Mp1p). Western blot analysis detected two yeast phase-specific immunodominant protein of *P. marneffei* (50KDa, 54 KDa).
6. *Molecular biology*: The PCR can detect *P. marneffei* targeting the genes for 5.8S rRNA and internally transcribed spacer (ITS1). The detection of 18S rRNA gene is possible by PCR hybridisation assay having sensitivity of 0.1 pgm/μL of DNA and one tube semi-nested PCR. A TaqMan real-time PCR is also developed to detect *P. marneffei* 5.8S rRNA gene in clinical samples and culture.

4.11.16 Treatment

P. marneffei is susceptible to miconazole, itraconazole, ketoconazole and flucytosine in vitro. The amphotericin B showed intermediate activity against the fungi.

4.12 *Cryptococcus*

Cryptococcus is a haploid-encapsulated yeast causing primarily central nervous system (CNS)-related disorders most commonly in dogs and cats and less commonly in ferrets, horses, cattle, goats, sheep and llamas and among nondomestic animals such as elk, koalas and dolphins. The yeast was first isolated from peach juice in Italy, and it was named as *Saccharomyces neoformans* (Sanfelice 1894) and also subsequently isolated from the infected tissues of a German patient (Busse 1894). Two years later, encapsulated bacilliform yeast, which was named *Saccharomyces subcutaneous tumefaciens*, was isolated from a healthy man in France, later identified as *Cryptococcus gattii* (Curtis 1896). Vuillemin examined several of these cultures, and due to lack of *Saccharomyces*-specific characteristics, he placed these species in the genus *Cryptococcus* in 1901. In 1905, von Hansemann reported the first case of meningoencephalic cryptococcosis. Emmons isolated *C. neoformans* from pigeon nests with their droppings and clinical cases of bovine mastitis (Emmons 1951, 1952). The first published report of *Cryptococcus* as an aetiology of disease in cats appeared in 1952 (Holzworth 1952). *Cryptococcus neoformans* var. *gattii* was formally proposed as a new taxonomic entity in 1970 (Gatti and Eeckels 1970). Ellis and Pfeiffer (1990) isolated *C. gattii* from the bark and fruits of *Eucalyptus camaldulensis* in Australia.

In India occurrence of *C. neoformans* was recorded from cutaneous lesion of cats (Pal and Mehrotra 1983b), mastitis in dairy cattle (Pal and Mehrotra 1983a) and meningitis in cats (Pal 1991).

4.12.1 Morphology

Cryptococcus life cycle is predominantly divided into two phases, i.e. vegetative and sexual

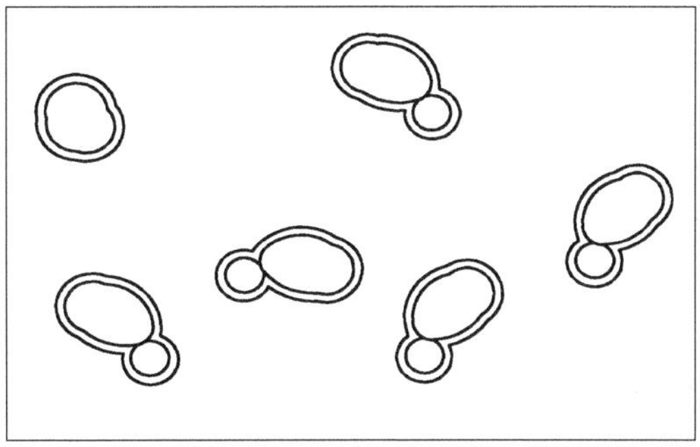

Fig. 4.25 Narrow-based buds of *Cryptococcus neoformans* (schematic)

growth phase. The sexual phase (teleomorph) of *C. neoformans* is known as *Filobasidiella neoformans*. Two major morphological forms exist in the vegetative growth phase. The predominant form found in the environment and animal hosts is unicellular budding yeast. The yeast cells are thin walled, spherical to oval with widely varying diameter (2–20 μm) and they are reproduced by mitotic division (Fig. 4.25). The buds are present in the narrow base. *C. gattii* yeast cells are oval and larger. Another vegetative form is pseudohyphae, produced by joined yeast cells due to incomplete separation after mitotic division. This form is sometimes found in the environment or the clinical samples produced to be protected from the natural predators.

The morphological transition from the yeast phase to hyphal phase is noticed during sexual mating. However, they are not considered as dimorphic fungi probably due to their predominant existence as yeast form in the environment and hosts and the lack of involvement of this transition in the pathogenesis. Further, both at 25 and 37 °C, they can produce yeast-like colonies in the isolation media. Recently unusually large yeast-like morphological form (30–100 μm) is also detected in clinical samples, known as 'giant' or 'titan' cells. These cells are produced as a response against host defence molecules, temperature, oxygen and CO_2, and the availability of specific macro- and micronutrients (e.g. glucose and iron). The G protein-coupled receptors (GPCRs) Ste3a and

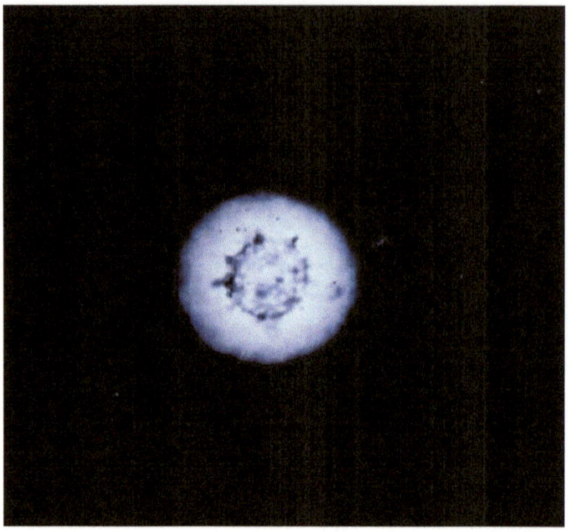

Fig. 4.26 Demonstration of *Cryptococcus neoformans* capsule in India ink stained smear (Photograph courtesy: Prof. P.P. Gupta, Ex-Director of Veterinary Research, Punjab Agricultural University, Punjab, India)

Gpr5 have been shown to regulate this gigantism of cryptococcal cells via the activation of G protein Gpa1 and its downstream signalling pathway (Gpa1-Pka1-Rim101signaling cascade). The phospholipids can also stimulate the giant cell formation.

Another unique morphological feature of *Cryptococcus* is the presence of capsule, like prokaryotes (Fig. 4.26). The capsule can be best observed in fresh preparations by staining with diluted India ink or phase contrast microscopy. Giemsa can also partially stain the capsule. In tissue section, it can be viewed with mucicarmine or Alcian blue stain. The capsule of *Cryptococcus neoformans* consists primarily of two polysaccharides, glucuronoxylomannan (GXM) and glucuronoxylomannogalactan (GXMGal), along with smaller amounts of mannoproteins. In addition to being associated with the cell, these molecules are also shed as an exopolysaccharide. It is noticed that the Ca^{++} acts as the bridge between the negatively charged glucuronic acid residues of neighbouring GXM fibres, mediating capsule assembly through GXM-GXM aggregation. So in absence of Ca^{++}, there is increased GXM shedding, and in the presence of EDTA the capsule size is decreased. The assembly of capsule is also influenced by the cell wall composition present beneath it. The cells without α (1, 3)-glucan in their cell walls do not display the surface capsule, whereas the cells lacking cell wall β (1, 3) glucan form larger capsules than those of wild-type cells. Two major transcription factors, namely, cryptococcal Nrg1 and Cir1, play central role in regulation of the capsule formation. Recently seven more transcriptional regulators that affect capsule size are also elucidated (Gat201, Tup1, Gcn5, Rim101, Hap3, Hap5 and Ada2). The capsule size is increased with iron deprivation, increased CO_2 concentration and the presence of serum in vitro and increased age of the organism or duration of infection in vivo.

Like other eukaryotic organism, *Cryptococcus* also possesses mitochondria which serves as a source of energy and is involved in different processes such as aging, calcium homeostasis, apoptosis and regulation of virulence (exclusive property of Cryptococcal mitochondria). Within the macrophages, the mitochondria may undergo fusion to generate 'tubular mitochondria' which are more effective in repair of mitochondrial DNA (mtDNA) damage caused by reactive oxidative species and increase the survival of the yeast. The organelle cannot be synthesised de novo; it is inherited directly from the single parent (*MATa*) during sexual reproduction probably due to migration of the nucleus from the *MATα* cell unidirectionally to the *MAT***a** cell, leaving behind its mitochondria. Whereas in other fungi [e.g. black ink mushroom (*Coprinus cinereus*)], this nuclear movement between two mating cells is bidirectional, resulting biparental mitochondrial DNA inheritance.

4.12.2 Classification

Cryptococcus belongs to the family Tremellaceae under the order Tremellales, class Tremellomycetes and phylum Basidiomycota. The genus *Cryptococcus* possesses over 37 species, majority of which do not cause disease in mammals. The most important pathogenic species are *C. neoformans–C. gattii* species complex, which includes *C. neoformans* var.

neoformans, *C. neoformans* var. *grubii* and *C. gattii* (*C. bacillisporus*). Sometimes interspecies hybrids of *C. neoformans* and *C. gattii* are observed. The teleomorphs of *C. neoformans* var. *neoformans* and *C. gattii* are *Filobasidiella neoformans* and *Filobasidiella bacillispora*, respectively. Other closely related species such as *Cryptococcus amylolentus*, *Tsuchiyaea wingfieldii* and *Filobasidiella depauperata* are isolated from insects.

4.12.3 Reproduction

The mating of *Cryptococcus* is not common in nature or within the host. In nature the proportion of two compatible yeast cells (α and **a**) are not equal which is the cause behind the absence of mating. The α-mating type predominates in nature and clinical specimens. Further, certain factors such as temperature (37 °C), high humidity and 5 % carbon dioxide (CO_2) can inhibit the mating which is commonly found within the host. In the laboratory, the specific stimulants (*myo*-inositol, indole acetic acid, copper ions and nitrogen starvation) can induce mating. The pigeon faeces can also favour the mating of *C. neoformans* var. *neoformans* and *Cryptococcus neoformans* var. *grubii* but not *C. gattii*.

During bisexual mating (opposite sex), two compatible yeast cells fuse to produce a hypha containing two parental nuclei (dikaryotic hypha/zygote). There will be a development of a specialised cell called basidium, in which the fusion of the parental nuclei and meiosis takes place. The daughter cells of the meiotic division undergo rounds of further mitotic division, and the mitotic nuclei are packaged into spores that bud from the apical surface of the basidium to form four spores in a basidium. After germination, the spores develop into new daughter yeast.

Same sex mating (unisexual mating/haploid fruiting/monokaryotic fruiting) is also detected in *Cryptococcus* generating monokaryotic hypha, basidium and spore. The unisexual mating involves fusion between two haploid cells of the same mating type, and it does not produce dikaryotic hyphae, and the spores are smaller and rounder. This type of mating generates diversity in *C. neoformans* populations and also it contributes to the global spread of cryptococcosis.

4.12.4 Susceptibility to Disinfectants

The replication of *Cryptococcus* stops above 40 °C. Alkaline pH is detrimental for the yeast. Among chemical disinfectants, *C. neoformans* is susceptible to 1 % sodium hypochlorite, iodine, phenolics, glutaraldehyde and formaldehyde.

4.12.5 Natural Habitat and Distribution

C. neoformans chiefly inhabits the avian excreta such as pigeon (*Columba livia*), parrot and canary excreta and in soil, decaying woods. The pigeon excreta are rich in creatine, urea and uric acid which act as major nitrogen source for the fungi. Transmission of the agent from the pigeon excreta to immunocompromised patients is also noted. Whereas the soil is the major reservoir of *C. gattii*, including the trees especially *Eucalyptus*, almond (*Terminalia catappa*), pottery trees (*Moquilea tomentosa*), etc. The decaying hollows in trunks and branches of a number of other tree species (>50 species) are also the preferred habitat of the fungus. However, *C. gattii* is generally not isolated from the pigeon faeces. The inositol produced by the plants can stimulate sexual reproduction of *Cryptococcus* which explains the utilisation of the environmental niche by the fungi. The organism was also isolated from the air, water bodies, car wheel wells and footwear sampled from high traffic location.

C. neoformans var. *grubii* is the most prevalent organism among *C. neoformans* group in human throughout the world. Human infections with *C. neoformans* var. *neoformans* are more common in France, Denmark and Italy. Previously it was observed that *C. gattii* was prevalent in tropical and subtropical regions such as Southeast Asia (New Guinea, Thailand), Australia, South America, and parts of Africa and the

United States (California). Currently the prevalence area is expanding throughout the world especially in North America.

In India, the reports of *C. neoformans* var. *neoformans* isolation from HIV patients and pigeon droppings in Tamil Nadu and *C. gattii* isolation from the trees (*Eucalyptus camaldulensis)* in Punjab are noticed. In cattle and buffalo, *C. neoformans* was detected from cases of mastitis in other states such as Madhya Pradesh. In cats, *C. neoformans* was isolated from meningitis and cutaneous lesion.

4.12.6 Genome

Cryptococcal genome is rich in introns (5.5 introns/gene) and antisense messages (endogenous antisense transcripts) in comparison to other Ascomycota yeasts. The reported genome sequence of two related strains of *C. neoformans* serotype D (JEC21 and B-3501A) revealed a 19-Mb genome sequence (excluding the ribosomal RNA repeats) which spans 14 chromosomes from 762 kb to 2.3 Mb. A total of 6572 protein-encoding genes were identified, which contain an average of 6.3 exons of 255 bp and 5.3 introns of 67 bp. There was no evidence of whole genome duplication. Overall 65 % of the genes of *C. neoformans* are conserved, 10 % of the genes are unique to *C. neoformans* and the remaining 25 % have non-fungal sequences. There are total 11 gene families, among them 2 are exclusive for *C. neoformans*, one encodes protein required for capsule formation and the other encodes nucleotide sugar epimerases associated with cell wall formation. Approximately 5 % of the genome is constituted with transposons clustered in single blocks (40–100Kb) that may represent sequence-independent regional centromeres. Each block contains Tcn5 and Tcn6 transposons. The transposons are also clustered adjacent to the rDNA repeats and within the mating-type (*MAT*) locus. The genome shows the evidence of alternative splicing (e.g. exon skipping, truncation, etc.) and antisense transcripts which have no coding potentiality. There is a single mating locus regulating the sexual growth phase spanning over 100 kb with more than 20 genes. It is quite larger than other *MAT* loci of different fungi. This locus can encode one of the two idiomorphic alleles which determines whether the cell will be *MATa* or *MATα*. In general two types of genes in a *MAT* loci, one encoding pheromone and pheromone receptor and the other encoding homeodomain transcription factor, are required for sexual mating. Initially during mating of *C. neoformans*, compatibility of the pheromone and pheromone receptor genes is required for cellular fusion, followed by dikaryon formation which is dependent on compatible homeodomain transcription factors (Sxi1α/Sxi2a). Further, *CLP1* gene products also have regulatory role in dikaryon formation which is Sxi-dependent. Mutation in Sxi1α gene can modify the mitochondrial DNA inheritance from uniparental to biparental.

Comparative genome analysis of *C. neoformans* var. *neoformans* and *C. neoformans* var. *grubii* revealed the transfer of a nonreciprocal 40Kb segment ('identity island') from *C. neoformans* var. *grubii* to *C. neoformans* var. *neoformans* nearly 2 million years ago. At present the 'identity island' is widely prevalent among the *C. neoformans* var. *neoformans* isolates in nature.

C. neoformans also contains small noncoding RNAs (sRNAs) like other eukaryotes such as small interfering RNAs (SiRNA) and micro RNAs (miR1 and miR2). The micro RNAs are 22 (miR1) and 18 nucleotides (miR2) in length and are derived from 70 nucleotide RNA precursors. In the genome, miR1 and miR2 are distributed in 4 (chromosome 1,4,9,12) and 7 chromosomes (chromosome 1,3,4,6,8,9,13), respectively. These sRNAs actively take part in gene silencing process via RNA interference mechanism.

The mitochondrial DNA (mtDNA) size varies from 24 to 34 Kb. The genome is present in high copy number and shows a higher mutation rate than the nuclear DNA. The genes associated with mitochondrial function (*ND1*, *ND2*, *ND3*, *ND4*, *ND4L*, *ND5*, *ND6*, *ATP6*, *ATP9*, *COX1*, *COX2* and *COB*) and protein synthesis (*SsrRNA* and

LsrRNA) are localised in the mitochondrial genome. The number of introns in some genes (*COX1*, *COB*, *LsrRNA*) vary between different species (*C. neoformans* var. *neoformans* and *C. neoformans var. grubii*) of *Cryptococcus*. In several introns of the *COB* and *COX1* genes, LAGLIDADG motifs are present. The intergenic regions are highly conserved in mtDNA.

4.12.7 Isolation, Growth and Colony Characteristics

Cryptococcus can be isolated in corn meal agar, Sabouraud dextrose agar, blood agar, honey agar, brain–heart infusion agar and malt agar. The specific medium is birdseed agar/Staib medium (niger seed agar)/sunflower seed extract agar with antibiotics to eliminate the bacterial contamination. They are sensitive to cycloheximide. The plates are incubated at 28–37 °C for 2 days–2 weeks. The pathogenic strains prefer to grow at 37 °C. The colonies are initially small, convex, mucoid, creamy in colour and increases in diameter up to several centimetres after prolonged incubation. In birdseed agar, the colonies appear as brown coloured in the centre of the plate due to the production of melanin. The optimum capsule production is detected in chocolate agar after incubation at 37 °C with 5 % CO_2 tension. The L-canavanine glycine bromothymol blue media can differentiate *C. neoformans* and *C. gattii* by the formation of distinctive blue colouration with the growth of *C. gattii*.

In *Cryptococcus* and a few other yeasts (*Candida albicans*), 'phenotypic switching' of the colony is observed. It is defined as spontaneous emergence of colonies with altered colony morphology at rates higher (10^{-2} to 10^{-5} per generation) than the somatic mutation (10^{-7} to 10^{-8} per generation). It is a reversible process and occurs only in a small fraction of the pathogenic population both in vitro and in vivo especially during chronic infection. It is detected in *Cryptococcus neoformans* serotype A, D and in *C. neoformans* var. *gattii*. The switching results the change of the smooth colonies into mucoid colony type. The smooth colonies are round with smooth-domed surface and smooth edges, whereas the mucoid colonies have a shiny mucoid surface due to excess production of viscous polysaccharide. These mucoid varieties show slower growth rate and produce the larger capsule. The major constituent of the capsule (GXM) is more viscous, can accumulate in the meninges and cerebrospinal fluid (CSF) and is more resistant to phagocytosis by the macrophages which enhance the virulence of the mucoid varieties. Further, the mucoid varieties can increase the intracranial pressure in chronic experimental cryptococcosis as observed in rat model.

4.12.8 Biochemical Characteristics

C. neoformans and *C. gattii* can hydrolyse urea (within 4 h) and assimilate inositol and creatinine, do not ferment carbohydrates and produce melanin which differ it from other pathogenic yeasts. They cannot reduce nitrate but can obtain their nitrogen from urea, creatinine and peptone. *C. gattii* can also use D-proline, glycine and tryptophan as a nitrogen source.

4.12.9 Antigenic Characteristics

The capsular polysachharide is the major antigen of *Cryptococcus*. The serotyping is based on the variations in the O-acetylation of the capsular polysachharide. *C. neoformans* was initially divided into five serotypes, A, B, C, D and AD. Currently, *C. neoformans* var. *neoformans* has the serotypes D, AD; *C. neoformans* var. *grubii* has the serotypes A, AD; and *C. gattii* has the serotype B, C. Further, molecular tools such as PCR-based fingerprinting, amplified fragment length polymorphisms, restriction fragment length polymorphism, random amplification of polymorphic DNA and multilocus sequence typing have generated several molecular types. *C. neoformans* var. *grubii* (serotype A) isolates belong to molecular types VNI and

Table 4.27 Serotypes and molecular types of *Cryptococcus*

	Cryptococcus neoformans var. *neoformans*	*Cryptococcus neoformans* var. *grubii*	*Cryptococcus gattii*
Serotype	D, AD	A, AD	B, C
Molecular type	VNIV, VNIII hybrid	VNI, VNII, VNIII hybrid	VGI, VGII, VGIII, VGIV

VNII, whereas *C. neoformans* var. *neoformans* is molecular type VNIV and the hybrid serotype AD is molecular type VNIII. *C. gattii* isolates are classified as VGI, VGII, VGIII and VGIV. The B and C serotypes of *C. gattii* are not specific to any molecular types (Table 4.27).

4.12.10 Virulence Factors

The virulence factors possessed by *C. neoformans* are described in Table 4.28.

4.12.11 Transmission

Cryptococcus is non-contagious, transmitted chiefly through inhalation or sometimes through percutaneous route in animals and human. The basidiospores, produced by sexual reproduction, are major infectious particles, small in size (2–3 μm) which can easily invade the lung alveoli than the encapsulated yeast (10–60 μm). The desiccated yeast form may become a source of infection, although they have limited viability in the environment. Rarely ingestion of large number of organisms may produce gastrointestinal lesion.

The children are often exposed to *C. neoformans* before the age of 5 years. In adults also antibodies to *Cryptococcus* is common, indicating their carrier state. In immunocompetent hosts, the organism is eliminated or it becomes dormant, whereas in immunocompromised hosts, they spread into other organs with fatal consequences. However, human-to-human transmission is not recorded, and it seems that human/animals are the dead-end host from where the cells can be recycled in the nature.

4.12.12 Pathogenesis

4.12.12.1 *C. neoformans* var. *neoformans*

The spores lodge in the lung alveoli of the animals or human after inhalation. Initially the infection is asymptomatic and dormant. The dormancy of the yeast cells is associated with several adaptive and immune evasion mechanisms. Calcineurin, produced by *C. neoformans* dephosphorylates a group of proteins that allow for their growth at physiological body temperature (37 °C), elevated CO_2 and alkaline pH. It also induces metabolic and oxidative stress genes required for their survival within the host tissues. The capsule of the fungi helps in evasion of immune system and survival within the host. The capsule is constituted with mannan (polysaccharide) which is highly hydrophilic. It makes a gelatinous zone surrounding the yeasts that conceals the pattern recognition receptors (PRR) from the immune system. Further, the mannan component can induce the production of IL10 (anti-inflammatory) and reduces the production of pro-inflammatory cytokines from the antigen-presenting cells and T lymphocytes. It also reduces the surface expression of activation markers and thus downregulates the macrophages, dendritic cells and neutrophils. The capsule also prevents antibody binding with the yeasts. Other capsular characteristics such as antiphagocytic property, shedding of L-selectin from the neutrophil surface associated with reduced adhesion and chemotaxis of neutrophils into the inflammatory site and altered phagocytosis (binding of capsular GXM with FcRγIIB of host cells produces IL10 which decreases the immune response) also help in immune evasion. However, persistence within the host for the longer period depends on survival of the yeasts within the macrophages and

Table 4.28 Virulence factors possessed by *Cryptococcus neoformans* var. *neoformans*

Virulence factors	Location in fungal cell	Function
Capsule	Outside the cell wall	(a) Prevention of phagocytosis by macrophages and neutrophils (b) Adherence with the host cell surface (c) Depletion of complements (d) Circulating capsular antigens help in the removal of selectins from endothelial cell surfaces and prevent the neutrophil migration into the tissues. Increased size is correlated with increased resistance against phagocytosis and antifungal therapy (e) Glucuronoxylomannan (GXM) component of the capsule interferes T cell activation and proliferation and disrupts cell-mediated immunity
Calcineurin (encoded by *CCN1* gene)		It helps the yeast to grow at body temperature (37 °C) and alkaline pH. The thermal tolerance is controlled by both calcineurin and the protein kinase C1 (PKC1)-activated MAP kinase (Mpk1) pathway
Melanin	Cell wall	It protects the organisms from oxidative damage by the scavenging host antioxidants, and it maintains the capsule structure or negative charge. Possession of melanin is also directly correlated with CNS invasion, as the dopamine is used as a substrate for melanin production It is catalysed by laccase enzyme which otherwise also decreases the hydroxyl radical level and offers protection to the organism. Both capsule formation and melanin production are controlled by Gpa1 G protein signalling pathway via regulation of cellular cAMP levels
Phospholipase B1 (Plb1)		The enzyme Plb1 is multifunctional containing the activities of phospholipase B, lysophospholipase and lysophospholipase transacylase. It helps in initiation and dissemination of *Cryptococcus* in the CNS and survival and replication within the macrophages. The Plb1 produces arachidonic acid from the host cellular phospholipid and the enzyme laccase produces prostaglandin E2 (PGE_2) from arachidonic acid The PGE_2 activates prostanoid EP2 receptor in the macrophages and inhibits microbicidal activity of the macrophages by increasing the cAMP level. The Plb1 mutants have reduced capability to escape macrophages (vomocytosis)
Urease		The enzyme urease can breakdown urea to generate ammonia which can produce a local damage in the endothelium, helping transmigration of the yeasts into the CNS
Superoxide dismutase [SOD1, SOD2 (mitochondrial)], glutathione peroxidases (GPX1 and GPX2), cytochrome C peroxidase (CCP1)		They are components of Cryptococcal antioxidant system which helps in the survival of the organism within the phagocytes
ATPase (encoded by *ENA1* gene)		It helps in the survival of the organism in extreme pH
acetoin and dihydroxyacetone	Cell metabolites	They inhibit the host neutrophil function
Inositol sphingolipid		Biosynthesis and breakdown of inositol sphingolipid by different enzymes are required for enhancing phagocytosis resistance and survival, invasion in CNS

(continued)

Table 4.28 (continued)

Virulence factors	Location in fungal cell	Function
CnRAM [Regulation of Ace2 and Morphogenesis (RAM) network]		Mutants lacking CnRAM pathway signal transduction showed slow growth at 37 °C and virulence
Vesicle ('virulence factor delivery bags')		It helps in carriage of virulence factors to the cell surface
Cir 1		It is GATA-type zinc finger protein which helps to maintain capsule size and integrity and enhance the growth at 37 °C. In addition, Cir 1 has negative control on melanin production with the help of Gat201. Further, they act as sensor for the iron level
Nrg1, Gat201, Tup1, Gcn5, Rim101, Hap3, Hap5, Ada2		They help to maintain capsule size and integrity, enhance the growth at 37 °C. Tup1, Nrg1 also act as transcriptional regulator to influence iron uptake
Sre1 [homolog of sterol-regulatory element-binding protein (SREBP)], Tup1, Vad 1		They help to maintain the melanin activity. Sre1 also regulates the genes encoding ergosterol biosynthetic enzymes and the proteins involved in iron and copper uptake, stress-related functions, various transport and metabolic functions under the low-oxygen condition (1 %). These measures are required for *Cryptococcus* to disseminate into the brain tissues, where oxygen tension is low than the atmosphere
Ferric reductase	Cell surface	They reduce ferric iron to its ferrous state which helps in iron uptake
Ferrooxide permease (Cfo1, Cft1)		They are essential for ferric iron uptake, iron acquisition from transferrin and full virulence in mice
Sit1		They are specific transporter of the siderophore ferrioxamine B. However, they are not required for virulence in mice model of cryptococcosis
Hap proteins (Hap3, Hap5, HapX)		They act as regulator for iron acquisition especially from the environment (HapX)
Copper regulatory transcription factor (Cuf1)		It controls melanin production, filamentation and growth at high temperature, dissemination to the brain (not in the lung)
Tubular mitochondria		It increases fungal survival within the macrophages
Dol-P-Man:protein *O*-mannosyltransferases (PMT)		This enzyme helps in *O*-mannosylation of the proteins in endoplasmic reticulum which is required for construction and maintenance of the cell wall. Loss of *pmt* gene can attenuate the virulence of *Cryptococcus*
cAMP-protein kinase A signalling (cAMP-PKA)		It is a signalling pathway involved in protein phosphorylation which is required for mating, capsule formation and melanin production
Histone acetyltransferase (Gcn5)		The *Gcn5* mutants show defective growth at high temperature, sensitive to oxidative stress and defective capsule attachment with the cell surface
Ubiquitin-conjugating proteins (encoded by *UBC6-2* and *UBC8*)		They are associated with oxidative stress response
F-Box protein (*fbp1*)		It is required for spore formation and the mutants are less virulent in experimental mice

endothelial cells. The survival strategy includes increased capsule synthesis, enzyme production (laccase, phospholipaseB1), production of melanin, Ssa1 (heat shock protein homolog), intervention with lipid metabolism, antioxidant system (superoxide dismutase, glutathione peroxidase), interference with nitric oxide synthase (NOS) induction leading to inhibition of nitric oxide production and capability to exit the cells (expulsion or phagosome extrusion). Further, *Cryptococcus* can modulate the adaptive immune response by several ways such as inhibition of T cell activation and induction of T cell apoptosis, induction of nonprotective Th2 response, interference with dendritic cell maturation and interference with antibody function that also helps in dormancy of the yeasts.

During immunosuppression of the hosts, the spores reactivate and disseminate into the different organs such as the skin, eyes, myocardium, bones, joints, lungs, prostate gland, urinary tract and the central nervous system (CNS). The dissemination into the blood circulation can occur through the infected macrophages ('Trojan horse mechanism'). They can multiply within the macrophages to produce 'cryptococcal phagosome' which lyses the macrophages to release the daughter yeast cells. Occasionally the daughter cells exit by extrusion without lysis. Sometimes they are again engulfed by fresh macrophages to avoid the contact with the host defence. After development of fungaemia, through the blood circulation, they can reach the blood–brain barrier (BBB) and penetrate the barrier to enter the CNS and cause rapid multiplication and meningoencephalitis as a consequence. The human or animal cerebellum is a rich source of inositol which maintains the normal neurological responses. *Cryptococcus* can use the inositol as a sole carbon source which explains their affinity for the brain.

Penetration of the barrier is crucial for the development of meningitis. They can induce the formation of microvilli-like protrusions by the reorganisation of the host cell cytoskeletal structures. These protrusions help the yeast cells to enter brain microvascular endothelial cells (BMECs). Sometimes the yeast cells utilise CD44 protein present in the lipid rafts of the host endothelial cells as receptor for invasion of BMECs (hyaluronic acid of the yeast acts as ligand). After close attachment with the host endothelial cells, the activation of host protein kinase C α-isoform occurs followed by cytoskeleton rearrangement that contributes to the intracellular transport. Within the cell cytoplasm, they are present in a membrane-bound vacuole, and they can exit from the cells with minimal damage. Thus, the endothelial cells are used as a vehicle by the yeasts to penetrate the barrier (transcytosis). Occasionally it is observed that occludin (tight junction maker protein) is degraded during the *Cryptococcus* and BMEC interaction which suggests the partial disruption of tight junction to allow the passage of the yeasts. Further, the 'Trojan horse mechanism' through the infected macrophages or other phagocytes is also used by the yeasts to penetrate the blood–brain barrier.

4.12.12.2 *C. neoformans* var. *gattii*

The pathogenesis described for *C. neoformans* var. *neoformans* is also applicable for *C. gattii* with some distinctions. *C. gattii* can arrest the neutrophil migration both in vivo and in vitro which explains the reason behind the susceptibility of healthy individuals to this fungus. Further it can produce extracellular fibrils which can prevent the phagocytosis by the neutrophils and help to establish the initial pulmonary infection. Trehalose is detected as major antioxidant in *C. gattii*. The genes encoding for trehalose synthase (TPS1p and TPS2p) are required for thermotolerance, virulence and expression of the capsule and melanin in *C. gattii*.

4.12.13 Disease Produced

The major animal and human diseases produced by different species of *Cryptococcus* are enlisted in Table 4.29.

4.12.14 Immunity

Innate immunity offers first line of defence against *Cryptococcus* by recognising the Cryptococcal patterns through toll-like receptor (TLR), β-glucan receptor, mannose receptor and the complement system. The professional phagocytes especially the bronchoalveolar macrophages engage in complement-mediated phagocytosis of the yeasts. The dendritic cells

Table 4.29 Major diseases of animal and human caused by *Cryptococcus*

Fungi	Host	Disease
Cryptococcus neoformans var. *neoformans*	Cats	It is the most common fungal infection in cats occurring in 3–7 years of age. The infection is characterised by formation of nasal proliferative lesion, granuloma in lymph node and the skin around the head and neck, fever, sneezing, head shaking and blindness. The fungi may invade the facial bone to cause distortion of nasal cavity. Sometimes the CNS involvement with meningitis is observed. Siamese cats are more susceptible than other breeds
	Dogs	Dogs 1–6 years old are susceptible. The infection is disseminated within the respiratory tract, ocular tissue, skin, lymph nodes, CNS. The subcutaneous granuloma surrounding the ears, face and feet is observed, although not so common like cats. The renal involvement is observed. The clinical signs include the respiratory signs (epistaxis, sneezing and nasal discharge) and neurological signs (stumbling, partial paralysis, ataxia, hyperesthesia, often along the dorsum or cervical area, severe seizure)
	Horse	Nasal granuloma, meningitis, granulomatous pneumonia, bowel or draining lymph node involvement (rare)
	Cattle	Mastitis with severe swelling and firmness of the udder, milk is mucoid and rarely the infection may be transmitted into the lung
	Human	In human the patients with suppressed immunity (HIV infected, suffering from lymphoma, haematological malignancy, using corticosteroid for prolonged period) are mostly susceptible to the fungal infection. It causes meningitis in human. One million cases of cryptococcal meningitis occur globally per year in AIDS patients, leading to approximately 6,25,000 deaths (majority of them in Africa)
Cryptococcus neoformans var. *grubii*	Cat	Infection in the nasal cavity
Cryptococcus gattii	Cats, dogs, goat, sheep, horse	The infection is restricted within the respiratory tract and central nervous system. The infection with VGII genotype involves several organs with grave prognosis, whereas infection with VGI genotype is restricted within the nasal cavity. The VGII genotype has hypervirulent strains
	Parrot	Mycotic rhinitis/lower respiratory tract infection, CNS involvement
	Pheasant	Enterohepatitis
	Pigeon	Opportunistic infection (lesion in face, lateral elbow, preen gland area and ankle, conjunctivitis)
	Human	Meningoencephalitis [in both the immunocompetent as well as immunocompromised patients (including AIDS patients)]

(DC), natural killer cells (NK) and neutrophils can also be lethal for the nonpathogenic *Cryptococcus*. The capsule can activate the alternative complement pathway and the binding of C3 with the yeasts. This C3-mediated binding is absent in CNS explaining another reason for their CNS affinity. Most pathogenic strains produce different virulence factors (Table 4.28) which can evade the innante response. The innate response helps to develop the adaptive immune response. The protective anti-Cryptococcal immunity is T cell-mediated (CD4+ Th1 and Th17). After effective clearance of the infection, regulatory T cell (Treg) response develops for the suppression of the response through the production of tumour growth factor (TGFβ) and IL10. However, pathogenic yeasts can also induce Treg response for their survival. The antibodies are also part of the immune response because the antibodies against the capsule component glucuronoxylomannan (GXM) can lyse the fungi in vitro.

The mucoid colony producers of *Cryptococcus* after switching generate a different type of immune response as observed in experimental pulmonary infection in mice. They produce more vigorous inflammatory response and a different cytokine profile than the smooth colony producers. They upregulate the production of monocyte chemoattractant protein (MCP) and macrophage inflammatory protein (MIP1α) chemokines and downregulate the production of IL10, IL4, IL2 and tumour necrosis factor (TNFα) which is correlated with massive lung damage and rapid death of the mice.

4.12.15 Diagnosis

4.12.15.1 Clinical Specimens

The samples include nasal swabs, tissue specimens collected by fine needle aspiration, blood, cerebrospinal fluid (CSF), urine, pleural and abdominal fluids, mastitic milk and lymph node, skin, lung and bone marrow (after postmortem).

4.12.15.2 Laboratory Examination

1. *Direct examination*: The smear can be prepared from the tissue samples and from the precipitate after centrifugation of the CSF collected from the suspected animals. The smear is stained with India ink or Nigrosin or Romanowski for demonstration of capsule. The Romanowski stain produces clearer capsule against the lightly stained background. The tissue biopsies can be stained with periodic acid–Schiff base (PAS) – haematoxylin stain – which will outline the yeast cell and the capsule will appear as clear zone surrounding the cell. In Mayer's mucicarmine stain, the capsule and cell wall appears as red.

   Another characteristic feature of *Cryptococcus* is narrow base budding (Fig. 4.25) in comparison to other yeasts (*Blastomyces*) having broad base budding (Fig. 4.10). Sometimes false-positive results are produced due to confusion with globules of myelin, lysed cells, lymphocytes and dead yeasts after successful treatment.
2. *Isolation and identification*: The centrifuged urine precipitate produces better isolation of the yeasts than the other clinical samples as most of the animals suffer from renal impairment. Media and incubation condition as described earlier will serve the purpose.
3. *Animal inoculation test*: Mice are inoculated intracerebrally or intraperitoneally or into the tail vein, and after 2 weeks they are sacrificed following the standard animal ethics protocol. In positive cases, gelatinous lesions are observed in the abdominal cavity and lungs. *Cryptococcus neoformans* var. *neoformans* is the only pathogenic species in mice. Rabbits are naturally resistant to cryptococcosis. This test is not considered as a standard practice nowadays.
4. *Detection of antigen*: Detection of Cryptococcal antigen in the CSF (antigenaemia titre) can be performed by latex particles coated with polyclonal serum and ELISA. Both of the tests are specific and sensitive. The detection of antigen is possible from both the live and dead organisms. During initial phase of therapy, disintegration of the yeast cells releases the capsule which produces high titre (>1: 60,000). So, the tests should not be done within 6–8 weeks after initiation of the therapy. The titre can be observed even after successful treatment as the dead organisms also have the intact capsule. However, negative titre is inconclusive because the relapse of the infection may occur if the organism persists in the nasal cavity. So, nasal culture should be done as a supplementary test for confirmation.
5. *Serological tests*: The serological tests for detection of Cryptococcal antibodies in serum are not very useful because the healthy animals also produce antibody titre due to carriage or persistent infection. Latex particle coated with monoclonal rabbit anticapsular antigen and indirect fluorescence antibody test (FAT) can be used for the detection of serum antibody. The false-positive reaction may occur in the patients infected with *Trichosporon*, *Mucor*, *Penicillium* and *Histoplasma*.

6. *Molecular biology*: Polymerase chain reaction (PCR) is developed to detect Cryptococcal *CAP 59* gene in biopsy samples collected from suspected animals.

4.12.16 Treatment

Treatment in the early phase of the infection in animals is successful. However, diagnosis without the laboratory help is difficult as most of the clinical signs are vague. Further, prolonged antifungal therapy (1–2 months) is required for effective clearance of the yeasts which may not be a cost-effective option for the owners. Amphotericin B (0.5 mcg/mL minimum inhibitory concentration) is the suitable antifungal for the animals especially with the CNS involvement. Other antifungals such as ketoconazole, itraconazole, albaconazole and fluconazole may be used. Itraconazole is more effective in the lung and bone involvement. Periodic assessment of liver enzymes is required with prolonged ketoconazole therapy in animals. Further, management of increased intracranial pressure is required especially during the first 3–4 weeks of the infection. Nonsteroidal anti-inflammatory drug is recommended as the corticosteroids are not useful in Cryptococcal infection. For evaluation, fungal antigen detection should be performed just after completion of therapy and 1 month later to know whether there is any relapse. The titre should be either undetectable or decreased by twofolds. Surgical intervention, cryotherapy and administration of amphotericin B and sodium iodide are recommended for treatment of Cryptococcal rhinitis.

In human, second-generation triazoles such as voriconazole and posaconazole are being used nowadays. The voriconazole has better penetration capacity of blood–brain barrier than posaconazole. However, antifungal therapy especially with amphotericin B can promote the selection of more virulent mucoid variety which often causes treatment failure.

The extracts of certain plants such as *Blepharispermum subsessile* (desmethyl isoencecalin), *Anacardium occidentale* (phenolic lipids, flavonoids, tannins), *Curtisia dentata* (triterpenoids), *Polygonum acuminatum* (sesquiterpenes) and *Daucus carota* (sesquiterpenes) have antifungal activity against *C. neoformans.*

4.13 *Candida*

Candida (previously known as *Monilia*) is imperfect unicellular dimorphic fungus which often reproduces by budding, and it can produce hyphae or pseudohyphae depending on the environmental condition. Hippocrates (460–377 BC) first documented oral pseudomembranous candidiasis, and he described it with the name of 'aphthae albae' which was later supported by Galen (130–200 BC). In modern era, Berg and Wilkinson separately described oral candidiasis (1841) and vaginal candidiasis (1849), respectively, for the first time. *Candida* was first described as a separate genus in 1923 by Berkhout. *Candida parapsilosis* was first isolated by Ashford from the stool of a diarrhoeic patient in Puerto Rico in 1928.

In India, Pathak and Singh (1962) reported for the first time the occurrence of *Candida albicans*, *C. tropicalis* and *C. paratropicalis* from the crops of the fowls. *Candida* was also isolated from empyema (Pal 1987), mastitis in buffaloes (Pal 1997) and pneumonia in goats (Pal and Lee 1999).

4.13.1 Morphology

There are three major morphological forms of *Candida* such as unicellular yeast, hyphae and pseudohyphae. The yeast form is oval (3.5–6 μm × 6–10 μm) with axial or bipolar budding. The hyphae are long tubes consisted of the cells with parallel sides, uniform width and true septum without any constriction (Figs. 4.27 and 4.28). The pores are present in septa for cell-to-cell communication.

The pseudohyphae are chains of elongated ellipsoidal cells with constriction between them. In the yeast and pseudohyphae, the nuclear division and septa formation takes place near the

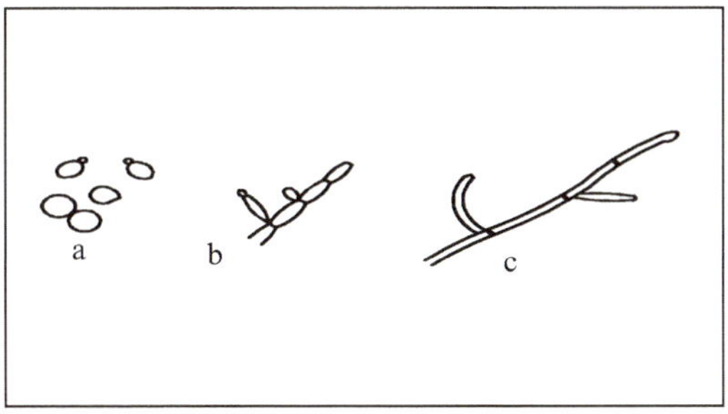

Fig. 4.27 Morphological forms of *Candida albicans* (schematic). *a* Yeast form, *b* pseudohyphae, *c* hyphae

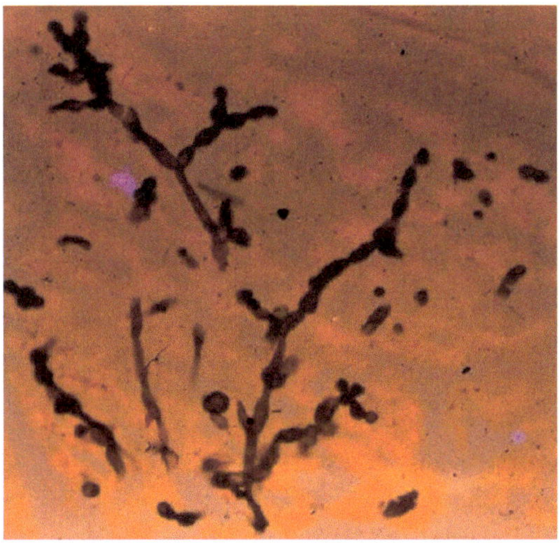

Fig. 4.28 Budding yeast, hyphae and pseudohyphae of *Candida albicans* in GMS stained udder tissue smear (Photograph courtesy: Prof. P.P. Gupta, Ex-Director of Veterinary Research, Punjab Agricultural University, Punjab, India)

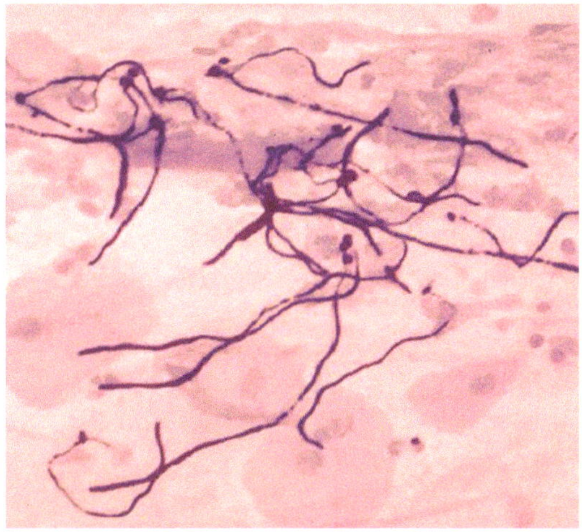

Fig. 4.29 Pulmonary candidiasis in Gram-stained sputum smear (Photograph courtesy: Prof. P.P. Gupta, Ex-Director of Veterinary Research, Punjab Agricultural University, Punjab, India)

bud, whereas in hyphae both the procedures occur within the *germtube*. Both the yeast and pseudohyphae can grow along with the cell cycle, but the hyphal cells remain arrested in G_1 phase after completion of first cell cycle until they can accumulate sufficient cytoplasmic mass to enter the second cycle. So, hyphae are less branched in comparison to pseudohyphae (Fig. 4.29). The hyphal cells contain a specialised organelle, known as *spitzenkorper*, which helps in growth at the hyphal tips. The intermediate morphological forms such as elongated single yeast cell, pseudohyphal cells with parallel sides and minor septa are also observed. The thick-walled chlamydospores remain attached with hyphae or pseudohyphae by a suspensor cell.

The cell wall is present outside the cell membrane composed of inner and outer layer. The inner layer is a meshwork of chitin, β-(1, 3) glucan and β-(1, 6) glucan which is more electron translucent. The outer layer is 150 nm width composed of mannoprotein. Three types of cell wall proteins (CWP) are present in the cell wall. The most abundant type is GPI-CWP which is covalently attached with β-(1, 6) glucan through glycophosphatidylinositol (GPI) anchor. Another class, known as Pir protein (*P*rotein with *i*nternal *r*epeats), is also covalently attached with β-(1, 3) glucan. The third category of proteins (Pra1p) lacks the covalent attachment and is heterogeneously distributed throughout the cell wall. Some of them are secreted into the environment.

The expression of morphological forms varies with the species of *Candida*. All the three forms are expressed by *C. albicans* and *C. tropicalis*, whereas other species such as *C. parapsilosis* can express the yeast and pseudohyphae forms. Some species such as *C. glabrata* can express the yeast and pseudohyphae forms, but primarily a yeast form both in the environment and host tissue without any morphological switch.

The morphological switch between the yeast and filamentous form is correlated with the virulence. Several external (environmental cues) and internal factors regulate the morphological switch of *Candida*. The environmental cues include presence of serum, temperature (37 °C), low levels of oxygen, high levels of CO_2 and poor nutrition. Bacterial peptidoglycan present in serum, *N*-acetylglucosamine found in mucus of gastrointestinal tract, 5 % CO_2 as a product of host cellular respiration and poor nutrition can induce the filamentous growth through the activation of cAMP/PKA pathway (regulated by *Candida* Ras protein). The pH change response is mediated through *Candida* Rim101 pathway. The contact-dependent response and osmotic sensing are regulated by Mkc1 and Hog1 MAPK pathway, respectively. The internal factors include the filament-induced gene (*HGC1*) which encodes a cyclin-related protein required for septin phosphorylation and inhibition of cell separation. Several members of secreted aspartyl proteinase (SAP) gene family (*SAP4*, *SAP5*, *SAP6*) are also expressed during morphological switch required for invasion of host tissues. The transition from yeast to hyphal phase in *Candida* is regulated by the activation of mitogen-activated protein (MAP) kinase and cAMP-protein kinase A (PKA) signal transduction pathways which can also coordinately regulate the virulence gene expression associated with this transition. Three sensor histidine kinases (Sln1, Chk1, Nik1) also regulate the phase conversion and mutants are unable to produce hyphae. Quorum-sensing (microbial communication) molecules such as farnesol, tyrosol and dodecanol are also associated with the transition.

Another type of morphological transition observed in *C. albicans* is known as white–opaque transition. In the laboratory, this switching is rare (1 in every 10^4 cell division) and it is regulated by host environmental factors. These morphological types are genetically stable, i.e. white or opaque cells always produce white or opaque type of progeny, respectively. The 'white' cells are relatively round and form smooth colonies on solid media, while 'opaque' cells are large and elongated and form flat and grey or opaque colonies. Certain phenotype-specific genes (*WH11*, *EFG1*) are expressed by white cells, while *OP4* and *SAP4* are expressed by opaque cells. White cells are more virulent and can easily colonise the host internal organs, whereas the opaque cells are associated with cutaneous infection probably due to expression of *SAP4* gene. The stable opaque type and either **a** or α-mating-type cell can undergo efficient sexual mating. Master regulator of this complex switching system is WOR1 which can convert the white cells into opaque cells. This WOR1 expression produces a direct positive feedback loop by binding its own promoter and turning 'on' its own expression. The a/α-heterodimer cells (mating type) cannot switch from white to opaque because they directly repress the WOR1 promoter. The host environmental factors regulating white–opaque transition are CO_2 and *N*-acetyl glucosamine (present in commensal bacterial cell wall and gastrointestinal tract mucus) which can produce stable opaque phenotype. Other factors, such as oxidative stress, UV light and adenine, also regulate white–opaque transition.

4.13.2 Classification

The genus *Candida* belongs to the class Saccharomycetes (Hemiascomycetes). It has more than 200 species. Approximately, 20 species have been associated with causing candidiasis in human and animals among which seven species are of major importance. *Candida albicans*, *Candida tropicalis* and *Candida glabrata* are the most frequently isolated from clinical specimens. The other pathogenic species are *C. parapsilosis*, *C. stellatoidea*, *C. guilliermondii*, *C. krusei* and *C. pseudotropicalis*. In dogs *C. natalensis* is the most frequently identified species.

4.13.3 Reproduction

Recent evidence suggests the presence of sexual reproduction in several *Candida* species especially in *C. albicans* (not yet detected in

C. parapsilosis and *C. tropicalis*). Two mating types (**a** and α) commonly exist in nature. For mating the cells have to be switched into opaque state. Once in opaque state, the **a-** or α-type of cells secrete the pheromones (MFa and Mfalpha) that are sensed via the cell surface receptors, Ste3 and Ste2, respectively. Pheromone signalling produces projections at the pole of the opposite type of cells (**a** and α). Subsequently fusion of the cells and karyogamy occur, producing mononuclear tetraploid cells. These tetraploid cells can undergo mitosis several times. In *Candida*, meiotic division is not observed to reduce the ploidy (formation of diploid from tetraploid cells). Instead, the concerted chromosome loss occurs to reduce the ploidy even in vivo, generating many aneuploid progeny cells. These cells have different filamentation, growth rate, white—opaque transition and increased antifungal resistance. The aneuploid cells with higher fitness are selected.

The a/α-heterodimer cells cannot directly participate in the mating as they cannot switch to opaque state. However, decreased expression of *HBR1* (haemoglobin response gene 1) causes the a/α-heterodimer cell to behave phenotypically like **a** cell. They are able to switch into the opaque state and participate in the mating.

C. albicans also shows the evidence of homothallic mating (**a-a** or α-α-mating) occurring between the same type of cells. The **a** cells can initiate the mating even if the α-cell is not available in the vicinity through autocrine pheromone signalling. The **a** cells release α-pheromone (Mfalpha) which binds with the Ste2 receptor either in the same or neighbouring **a** cells. It is followed by **a–a** fusion of the cells. High level of Bar1 protein can inhibit the homothallic mating between the opaque **a** cells.

4.13.4 Susceptibility to Disinfectants

Crystal violet is an effective disinfectant against *C. albicans*. Among antiseptics, 4 % chlorhexidine gluconate in alcohol and 10 % povidone–iodine have potential antifungal activity against *Candida*, whereas among disinfectants, quaternary ammonium compounds (1:10,000) is lethal at short contact time.

Among the physical agents, *Candida* is susceptible to heat (more than 50 °C) and ultraviolet light and resistant to freezing.

4.13.5 Natural Habitat and Distribution

Candida normally inhabits in the mucosal layer of animals and human such as alimentary, upper respiratory, genital tract and oral mucosa such as posterior dorsum of tongue. They can invade the deeper part of the tissues to establish the infection under immunosuppressed conditions caused by prolonged antibiotic or steroid use, inflammation and breakage of epidermal layer by injury. From the environment, *Candida* is commonly isolated even from hypersaline niche.

C. albicans is distributed throughout the world.

4.13.6 Genome

Each *Candida* species contains 6–9 numbers of chromosomes. The genome is diploid in most of pathogenic *Candida* species such as *C. albicans*, *C. tropicalis* and *C. parapsilosis*. The haploid genome is detected in C. *guilliermondii* and *C. lusitaniae*. The size of *C. albicans* genome is 10.6–15.5 Mbp with the number of genes varying from 5,733 to 6,318. The genome contains the major repeat sequence (MRS) elements where the genetic recombination can take place. Another feature of *C. albicans* genome is the presence of one single nucleotide polymorphism (SNP) in every 330 bp (SC5314 strain) to 390 bp (clinical strain WO-1).

At the transcription level, the sexual mating of *Candida* is regulated by *MTL* (mating-type locus) loci present in the genome. The transcription factors a2, α1 and a1/α2 are required for the expression of specific genes and generation of **a-**, α- and a/α-heterodimer cells. The locus also contains the additional genes such as gene for phosphatidylinositol kinase (*PIK*), oxysterol-binding proteins (*OBP*) and poly A polymerases (*PAP*) with unknown role in mating. In some species of *Candida*, homothallic mating between same sexes is detected where either the *MTL* is absent or both the loci are fused into a single locus.

4.13.7 Isolation, Growth and Colony Characteristics

Candida can be isolated in general fungal or bacteriological media such as Sabouraud dextrose agar (with penicillin, streptomycin, chloramphenicol), potato dextrose agar, blood agar and brain–heart infusion agar. They are obligate aerobe and they can grow within a wide range of temperature and pH. The plates are incubated at 25–30 °C for 2–3 days. Pigmentation is detected in corn meal tween agar. Currently the media with resins are in use which can absorb the residual antifungal or other inhibitory substances present in the clinical samples and significantly improves the recovery of *Candida*. Further, *C. albicans* can grow in the presence of 0.04 % cycloheximide.

The colonies are circular, white or opaque in colour (white–opaque transition is required) and creamy in consistency.

For identification of *Candida* at species level, a chromogenic medium (CHROMagar) can be used in which different species such as *C. albicans*, *C. tropicalis* and *C. krusei* produce green/bluish-green, blue/purple with halo and pink-coloured/ruffled colonies, respectively. Similarly, Oxoid Chromogenic *Candida* Agar (OCCA) can also differentiate the species with the colour of the colonies. The sunflower seed agar (Pal's medium) can differentiate between *C. albicans* and *C. dubliniensis* where only *C. dubliniensis* isolates exhibited a hyphal fringe surrounding the colonies.

4.13.8 Biochemical Characteristics

Production of urease and carbohydrate fermentation pattern (Table 4.30) helps in identification of *Candida* species.

4.13.9 Antigenic Characteristics

Candida has two major groups of antigens such as heat-stable polysaccharides (glucan, mannan) and heat-labile glycoproteins. The mannan is a major antigen of *Candida* that circulates during infection and it comprises 7 % dry weight of the cell wall. It is resistant to heat, proteinase and acidic pH. The antigenicity varies with the length of the polysaccharide side chain and position of the glycosidic linkages (α-man and β-man indicating the α-/β-linked oligomannose residues). *C. albicans* has two serotypes (A and B) based on these variations.

4.13.10 Virulence Factors

The virulence factors possessed by *C. albicans* are described in Table 4.31.

4.13.11 Transmission

Candida causes the opportunistic infection in most of the clinical cases during immunosuppression. In bovine udder, they may be transmitted by the contaminated milker's hands during milking or administration of medicines.

Table 4.30 Carbohydrate assimilation pattern of different *Candida* species

Carbohydrate	*C. albicans*	*C. tropicalis*	*C. pseudotropicalis*	*C. krusei*	*C. parapsilosis*
Glucose	+ (acid and gas)	+ (acid and gas)	+ (acid and gas)	+ (acid and gas)	+ (acid and gas)
Maltose	+ (acid and gas)	+ (acid and gas)	–	–	–
Sucrose	+ (acid only)	+ (acid and gas)	+ (acid and gas)	–	–
Lactose	–	–	+ (acid and gas)	–	–

Table 4.31 Virulence mechanisms and factors possessed by *C. albicans*

Virulence factors	Functions
Yeast–hyphae transition	The hyphae are more invasive due to application of mechanical force and can easily penetrate both the individual cells and the space between them, whereas the yeast cells are easily disseminated into the blood circulation. The *Candida* hyphae can damage endothelial cells, lyses the macrophages after phagocytosis
Thigmotropism (directional hyphal growth through contact sensing)	It is the contact sensing mechanism of *Candida* hyphae by which they can identify the crevices, grooves present in the host tissues for effective penetration. The contact sensing also triggers the biofilm formation over the solid surface. It is regulated by extracellular calcium uptake through the calcium channels Cch1, Mid1
Biofilm	*Candida* is able to produce biofilm in host mucosal cell surfaces and medical devices along with some bacterial species. Within biofilm the cells have reduced growth rate due to limited nutrition. The *CDR* and *MDR* genes encoding two types of efflux pumps, i.e. ATP-binding cassette (ABC) transporters and major facilitators, are upregulated in biofilm cells which are associated with antifungal resistance along with other factors such as complex architecture and metabolic plasticity
B cell mitogenic protein (ISM p43)	It causes hyper stimulation of B cells and immunosuppression and helps in survival of the fungi within the host
HYR1 (hyphal protein)	It is associated with antifungal resistance
pH-sensitive protein (PHR1 and PHR2)	They are cell wall β-glycosidases. The PHR1 and PHR2 are expressed in neutral-alkaline and acidic pH, respectively. So they are associated with systemic and vaginal infection, respectively
Extracellular pH modulation	*Candida* can modulate the extracellular pH towards alkaline and auto-induce the hyphae formation. It is associated with starvation and uptake of amino acids, polyamine, etc. The amino acids are cleaved intracellularly by fungal urea amidolyase (Dur1,2) to generate ammonia which is exported into the external environment through the Ato (**a**mmonia **t**ransport **o**utward) export proteins. The export of ammonia makes the environment alkaline which promotes the hyphal growth
Metabolic plasticity	In the hostile environment (within macrophage), *Candida* promptly switches glycolysis to gluconeogenesis where lipid and amino acids serve as a nutrient source
Iron acquisition	*Candida* can acquire iron through different mechanisms such as reduction of host ferritin (Als3 mediated), acquisition of iron from the siderophores produced by other organisms (xeno-siderophore) and uptake from haemoglobin and other haeme proteins
Zinc acquisition	It is mediated by zinc-binding protein (Pra1: **p**H-**r**egulated **a**ntigen 1) which acts as a zincophore by binding extracellular zinc
Enzymes	
Secreted aspartyl proteinases (Sap 1–8; Sap 9,10 cell bound)	The enzyme degrades immunoglobulin and complement of the hosts. Its help in penetration of host tissues is controversial as detected in the recent studies
Phospholipase A, B, C, D; lysophospholipase and lysophospholipase-transacetylase	They are associated with host cell membrane damage (PLB), adherence and penetration
Haemolysin (mannoprotein)	They lyse the erythrocytes to release iron required for survival of the fungi
Catalase and superoxide dismutase	It provides protection against oxidative stress
Adhesins/invasins	
Glycolytic proteins (GAPDH, PGK, ADH, enolase), heat shock proteins (Hsp104, Hsp90, Hsp78, Hsp70, Hsp60) Small Heat shock proteins (sHsp) [Hsp31, Hsp30, Hsp21, Hsp12, Hsp10]	(a) They act as adhesion factors and modulators of host antifungal immune response (b) The heat shock proteins can further act as chaperones and prevent protein unfolding and aggregation under stress such as high temperature, starvation and oxidative stress (c) The sHsp proteins are low molecular weight chaperones that prevent protein aggregation

(continued)

Table 4.31 (continued)

Virulence factors	Functions
Lectin-like molecules	They help in adhesion of *Candida* with L-fucose, *N*-acetylgalactosamine or *N*-acetylglucosamine containing glycosides present in the epithelial cell receptors. They also recognise salivary protein (mucin) and bacterial cell surface (*Staphylococcus*) which helps in oral colonisation
Fimbriae	It helps in the adhesion of *Candida* with the erythrocytes mediated through the fimbrial protein (66 Kda) and glycosphingolipids present in the erythrocyte surface
Integrin analogue [a protein that can bind RGD (arginine-glycine-aspartic acid) motif]	(a) They help in the adhesion of *Candida* with the proteins having RGD motifs such as extracellular matrix protein, complement (C3), fibrinogen, heterodimeric transmembrane proteins (b) They play major role in yeast–hyphae transition.
Mannan	They help in adhesion of *Candida* with epithelial and endothelial surfaces, erythrocytes, salivary components
ALS protein [encoded by *ALS* gene (agglutinin-like sequence)] consisting of eight members (ALS 1–7, ALS 9)	The hyphae-associated ALS protein (ALS3) adheres with host laminin, fibronectin and collagen through the recognition of certain amino acids (threonine, serine or alanine). The ALS3 also acts as invasin and is associated with induced endocytosis by the host epithelial cells
Hwp1 (a hypha-associated GPI-linked protein)	It acts as a substrate for mammalian transglutaminases and this reaction may covalently link *Candida* hyphae to host cells
Ssa1 (cell surface-expressed protein belonged to HSP70 family)	It acts as invasin and is associated with induced endocytosis by the host epithelial cells
Fibronectin adhesin	They help in the adhesion of *Candida* with fibronectin and vitronectin receptors
Flucoside-binding adhesin	They help in the adhesion of *Candida* with glycoside receptor

4.13.12 Pathogenesis

4.13.12.1 *C. albicans*

Adhesion to the host epithelial cells is a prerequisite step for both commensal and pathogens. The adhesion is a complex process as both the cells (*Candida* and host cells) are negatively charged which will repeal each other. To overcome this repulsive force, some other attractive forces also act such as Van der Waals force, hydrophobic interaction and Brownian movement force. Successful adherence depends on the interaction between these forces [extended Derjaguin–Landau–Verwey–Overbeek (DLVO) theory]. Overcoming the repulsive forces, *C. albicans* detects the host environment either by contact sensing due to body temperature, pH change or any other unexplored mechanism to express different invasins (Table 4.31) on their surface for attachment with the host cells.

Majority of the *C. albicans* strains with hidden pathogen-associated molecular pattern (PAMP) can adapt the epithelial surface environment and remain as commensal without any cell damage and further cytokine response a few strains express the virulence factors which also depend on host conditions such as immunosuppression due to prolonged use of glucocorticoid, breaching of the mucosal surface, concurrent infection, neutropenia associated with chemotherapy, etc. These strains are either eradicated by the host immune system or they become able to establish a systemic infection.

Following adhesion, *C. albicans* changes morphologically from the yeast phase to the hyphal phase required for tissue penetration. They use two different mechanisms such as 'induced endocytosis' or 'active penetration' for invasion into the host epithelial cells. For induced endocytosis, the fungi express different proteins in their surface ('invasins' such as ALS3, Ssa1) which can bind the ligands present in the host cells (E-cadherin on epithelial cells and N-cadherin on endothelial cells). This binding stimulates the host cells for engulfment of the hyphae (both live and dead) by a clathrin-dependent mechanism or macropinocytosis. Active penetration is conducted by live hyphae only with the help of mechanical force and damaging factors which can disrupt the host cell membrane, although the detailed mechanism is still unknown. The damaged epithelial cells in turn stimulate the release of pro-inflammatory cytokines which attract neutrophils and macrophages in the area for destruction of the invading hyphae.

Within the macrophages, *Candida* uses several survival strategies such as production of catalase and superoxide dismutase, providing protection against oxidative stress. *C. albicans* also delays phagosome maturation and induces recycling of late maturation markers like LAMP-1. Soon the hyphae disrupt the macrophage membrane and escape. Even recently non-lytic escape mechanism of *C. albicans* is also detected as observed in *Cryptococcus*. However, *C. glabrata* not only survives, but replicates inside the phagosome until the phagocyte bursts due to the modification of normal phagosome maturation process. Their phagosome lacks any lysosome marker such as cathepsin D and remains non-acidified.

For the establishment of infection, *C. albicans* requires macro- and micronutrients (iron, zinc) from the host. They can metabolise different sugars and can use all amino acids as nitrogen sources. In addition, *C. albicans* possesses several families of secreted hydrolases and transmembrane transporters (Table 4.31) which help in the macro- and micronutrient acquisition. During disseminated candidiasis, *C. albicans* gains access to the bloodstream, relatively rich in glucose (6–8 mM) which is the preferred nutrient source of most fungi.

4.13.13 Disease Produced

The major animal and human diseases produced by different species of *Candida* are enlisted in Table 4.32.

4.13.14 Immunity

The innate immune system is activated through the recognition of Candidal pathogen-associated molecular pattern (PAMP) by the pattern recognition receptors (PRR) expressed by different immune cells (Table 4.33).

The PAMP molecules are present both in outer and inner layer of *Candida* which is exposed during budding or during exposure with f antifungals and host enzymes. Recognising the specific Candidal PAMP, the TLR4 induces pro-inflammatory signals in monocytes, macrophages and dendritic cells (DCs) through the MyD88-Mal-mediated NFκB and mitogen-activated protein kinase (MAPK) pathways. They also stimulate TH1 responses through IRF3-dependent mechanisms which provide protection against disseminated Candidiasis, whereas the TLR2 stimulates the production of moderate amounts of pro-inflammatory cytokines and strong IL10 and TGFβ responses. The dectin1 induces IL2, IL10 and TH17 responses through a Syk-CARD9 cascade and increases the effects of TLR2. The mannose receptor (MR) induces pro-inflammatory effects in monocytes and macrophages. Other pathways include production of tumour necrosis factor (TNF) and IL1Ra by dectin 2 and production of the immunosuppressive cytokine (IL10) through DC-SIGN.

The mucosal surface of healthy individuals is often colonised with *C. albicans*. The mucosal cells provide a structural barrier and they can recognise the fungi. The epithelial cells can respond to *C. albicans* in a two-phase MAP kinase (MAPK) pathway that enables them to discriminate between the commensal yeast and the invasive pathogenic form. The second-phase MAP kinase (MAPK) pathway can be induced by the invasive hyphae, not the yeast. This second-phase induction leads to activation of the transcription factor cFos and the MAPK–MKP1 which is necessary for the production of cytokines (IL1, IL6). The hyphae can activate

Table 4.32 Major diseases of animal and human caused by *Candida* sp.

Fungi	Host	Disease
Candida albicans	Poultry	*Thrush (crop mycosis)*: It is observed in chicken, turkey, guinea fowls. In acute cases, the birds may die without any syndrome. The crop, mouth, oesophagus, proventriculus and gizzards are commonly affected. The lesions are white plaque, ulcer and/or pseudomembrane adherent with the mucosal surface. The crop mucosa produces characteristic 'Turkish towel' appearance. The disease is associated with coccidiosis, overcrowding
Candida parapsilosis, *C. krusei*, *C. tropicalis*, *C. rugosa*	Cattle	*Mycotic mastitis*: Usually prevalence is low in cattle herd (10 %) and the infection is mild and self-limiting in nature. It has been associated with using contaminated syringes, canulars and antibiotic preparations. Teat injuries may act as predisposing cause. It is transmitted either from contaminated udder skin through the milker's hand, milking machine or from the contaminated floors, straw, feed, sanitary agents and other equipments
Candida albicans, *C. krusei*	Foals, calves	Gastroesophageal candidiasis and gastric ulceration
Candida albicans	Pigs	Alopecia, circular skin lesion covered with exudates, pseudomembrane formation
Candida albicans, *C. parapsilosis*, *C. guilliermondii*	Dogs	Exfoliative dermatitis in the muzzle, scrotum and feet along with pruritis and alopecia, otitis externa, genital tract infection
C. albicans, *C. glabrata*	Dogs	Canine peritonitis with marked pyogranulomatous inflammation
Candida albicans, *C. parapsilosis*	Cats	Localised and disseminated infection associated with feline panleukopenia and feline immunodeficiency virus, granulomatous rhinitis associated with chronic use of immunosuppressive drug, enteritis
Candida albicans, *C. glabrata*, *C. parapsilosis*, *C. tropicalis*	Human	(a) *Invasive candidiasis*: It is associated with immunosuppression, neutropenia due to stem cell or solid organ such as liver transplantation. *Candida* is among the important pathogens in the patients admitted to the intensive care unit. It causes high rate of morbidity and mortality without specific syndrome (b) *Oral candidiasis*: The patients receiving broad spectrum antibiotics, steroid or cytotoxic therapy and having diabetes mellitus, xerostomia, nutritional deficiency and AIDS are susceptible to oral candidiasis. It is characterised by white patches on the surface of oral mucosa (thrush); erythematous lesion on the dorsum of the tongue, palate or mucosa (antibiotic sore mouth); formation of leukoplakia (chronic lesion); denture-induced stomatitis; glossitis; etc.

the inflammasome and induce IL1β production in immune cells such as macrophages and stimulate TH17 cells to produce cytokines (IL17, IL22). The IL17 activates neutrophils, while IL22 induces the release of defensins from epithelial cells; both are crucial components of mucosal antifungal defence.

4.13.15 Diagnosis

4.13.15.1 Clinical Specimen

The clinical specimens include skin scrapings, centrifuged milk samples, blood, serum and tissue samples in 10 % formalin for biopsy.

4.13.15.2 Laboratory Examination

1. *Direct examination*: The smear prepared from the clinical specimens can be observed under microscope either by 10 % KOH preparation or by staining with Gram stain method. In the tissue section, *Candida* can be observed by PAS-haematoxylin or methenamine silver stain. They appear as unicellular budding yeast or pseudohyphae.
2. *Isolation and identification*: Media and incubation condition as described earlier will serve the purpose.
3. *Demonstration of germ tube formation*: A single Candidal colony is picked up and mixed with 0.5 ml of rabbit or sheep or bovine or

Table 4.33 Pathogen-associated molecular pattern (PAMP) of *C. albicans* and their pattern recognition receptors (PRR) expressed by different immune cells

PRR	Immune cells	PAMP
Toll-like receptors (TLR)		
TLR2	Macrophage, neutrophil, dendritic cells, CD4+ T cells	Phospholipomannan
TLR4	Macrophage, neutrophil, dendritic cells, CD4+ T cells	Mannan
TLR9	Macrophage, neutrophil, dendritic cells	CpG oligodeoxynucleotides
C-type lectin receptors (CLR)		
Mannose receptor	Macrophage, dendritic cells	Mannan
Dectin-1	Macrophage, neutrophil, dendritic cells	β-1,3-glucan
Dectin-2	Macrophage, neutrophil, dendritic cells	High-mannose structures (Man9GlcNAc2)
DC-SIGN	Macrophage, dendritic cells	High-mannose structures (Man9GlcNAc2)
Galectin-3	Macrophage	β-1,2-mannosides
Mincle (Clec4e and Clecsf9)	Macrophage	Unknown
Complement receptor 3 (CR3)	Neutrophils	β-glucan
Langerin	Langerhans cells	Mannose, fucose
NOD-like receptors (NLR)		
NLRP3 (NLR family pyrin domain containing 3)		Unknown

human serum and incubated at 37 °C for 1–2 h. After the incubation, a drop of preparation is observed under the phase contrast or dry objective of the microscope. In positive case, small tubes projecting from the yeast cells without any constriction at the point of attachment are observed. This is a characteristic of *C. albicans* and *C. tropicalis* (after prolonged incubation for 3 h).

4. *Demonstration of chlamydospore production*: For demonstration of chlamydospore, single isolated Candidal colony can be inoculated into the cornmeal tween 80 agar or chlamydospore agar by making the parallel cuts in the media. Sometimes below the surface inoculation is preferred for better chlamydospore production in low-oxygen tension. The plates are incubated at 30 °C for 2–4 days. A coverslip is placed over the surface, and the plate is observed under the high dry objective for the detection of thick-walled chlamydospores (8–12 μm) at the tips of pseudohyphae.
5. *Histopathology*: The presence of blastospores and pseudohyphae in the histochemically stained smear can be identified. A monoclonal antibody (3H8) is developed against *C. albicans* cell wall mannoprotein which can specifically recognise *C. albicans* in culture and in paraffin-embedded tissue sections using immunofluorescent and immunohistochemical staining. This antibody specifically detects the mycelial forms and to a lesser extent blastospores of *C. albicans* and does not react with any other *Candida* spp. Fluorescent in situ hybridisation (FISH) using oligonucleotide probes directed against 18S rRNA has been used to differentiate *C. albicans* from *C. parapsilosis* in tissues.
6. *Immunological assays*: The ELISA-based tests are developed to detect the Candidal antigens such as secreted aspartyl proteinase, mannan and β-D glucan. Dissociation of antigen–antibody complexes is necessary for the optimal detection of mannan in the circulation. This antigen is heat stable and resists boiling, proteinase treatment and acidic pH. So antigen–antibody complexes are routinely dissociated by boiling with EDTA or by

enzymatic treatment. The sensitivity and specificity of the mannan-based test is 58 and 93 %, respectively, whereas the sensitivity and specificity of β-D glucan (BDG) detection test is 77 and 85 %, respectively. These immunological assays are useful in diagnosis of *Candida* in immunocompromised patients who produce negligible or undetectable amount of antibodies.

7. *Serological tests*: The ELISA-based serological tests are developed for detection of antibodies against Candidal aminopeptidase, mannan and secreted aspartyl proteinase in the suspected serum. Indirect immunofluorescence assay is also developed for the detection of anti-Candidal antibody (IgG). However, serological tests for detection of *Candida* are not so much reliable. The false-positive reaction occurs due to superficial colonisation, and false-negative reaction occurs in immunocompromised patients producing low or undetectable level of antibodies.
8. *Molecular biology*: Several types of PCR such as semi-nested and nested PCR, multiplex PCR followed by DNA sequencing or pyrosequencing are developed for detection of Candidal DNA. The target DNA includes rRNA (5.8S, 18S, 28S) gene, internal-transcribed spacer (ITS) and intergenic spacer (IGS) region. However, liberation of DNA from the cell is challenging due to the presence of a highly rigid cell wall which often produces false-negative result. The false-positive result is produced when highly conserved rRNA (5.8S, 18S, 28S) or other genes are targeted and airborne fungal contamination occurs. Confirmation of PCR amplicon by enzyme immune assay (EIA) using colorimetric substrate is considered as an easy and cheapest method.

   Real-time PCR method is also developed using Taqman probe or light cycler system for the identification of *Candida*. They also lack the post amplification detection costs; however, the initial cost for setup of the laboratory and instruments is high.

4.13.16 Treatment

Copper sulphate in drinking water or nystatin in feed is the traditional treatment of crop mycosis in poultry. Currently amphotericin B and fluconazole are also in use. The amphotericin B is effective against wide species range of *Candida* and is relatively cheap. However, nephrotoxicity is detected as side effect in human patients.

Experimentally Linum usitatissimum (linseed/flaxseed)-fixed oil and the herb (*Persicaria senegalense*) are found effective against cattle mastitis causing strains of *C. albicans*.

The cutaneous infections in animals are treated by removal of crusts which is followed by topical application of povidone–iodine, copper sulphate or chlorhexidine or selenium sulphide in the area.

Other antifungals such as voriconazole, caspofungin and micafungin are also effective against *Candida* with minimal side effects. They are either expensive or their veterinary preparation is not always available.

4.14 *Sporothrix*

Sporotrichosis was first described by Linck in 1809 and Lutz in 1889, but they could not isolate the fungi. In 1896, *Sporothrix schenckii* was first isolated by Benjamin Schenck, a medical student at Johns Hopkins Hospital in Baltimore, USA. He isolated the organism from a male patient with cutaneous lesion in the right arm. The isolate was confirmed by Erwin Smith, a mycologist who described the fungus as *Sporotrichum*. In 1900, Hektoen and Perkins further reported sporotrichosis from cutaneous lesion of a patient in the Unites States. The dimorphism of *Sporothrix* was described by Howard in 1961. However, the erroneous classification of the fungi under the genus *Sporotrichum* continued until 1961. Carmichael (1962) first recognised the differences of the *Sporothrix* isolates from the *Sporotrichum* genus, and finally the nomenclature *Sporothrix schenckii* was accepted.

In 1907, the first natural animal infection with *Sporothrix schenckii* was reported by Lutz and Splendore in rats from Brazil.

In India, the first authentic report of human sporotrichosis was published in 1986. The isolate was obtained from a 55-year-old housewife with multiple lymphocutaneous lesions on her arm, residing at Ranikhet, Uttar Pradesh. Further human sporotrichosis was reported from Himachal Pradesh, Manipur, Uttarakhand and Bangalore (Sharma et al. 1990). In 2009, feline transmission of lymphocutaneous sporotrichosis in human was also reported from Karnataka, India (Yegneswaran et al. 2009).

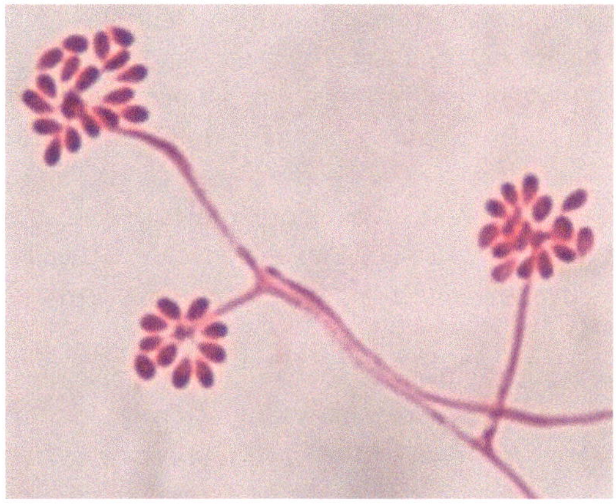

Fig. 4.30 Daisy flower appearance of *Sporothrix schenckii* (Photograph courtesy- Prof. Alexandro Bonifaz, Head, Department of Mycology and Dermatology service, General Hospital of Mexico, Mexico City)

4.14.1 Morphology

Sporothrix schenckii is a dimorphic fungus with a saprophytic mycelial and parasitic yeast phase. The mycelium is composed of slender (1–3 μm), hyaline, septate and branched hyphae bearing thin conidiophores. The apex of conidiophore carries a small vesicle with symmetrically arranged denticles. Each denticle bears one conidium. Each of the conidium is pear shaped to clavate, approximately 2–4 μm in measurement and do not yield any chain. These conidia are arranged like the petals of flower forming a 'peach flower' or 'daisy flower' appearance (Fig. 4.30). Occasionally, dark, thick-walled conidia are directly attached with the hyphae which are arranged bilaterally side by side in a row (aleurioconidia or raduloconidia). These dark cell-walled conidia are specific for *Sporothrix schenckii* and are not detected in nonpathogenic species. The yeast cells are pleomorphic showing oval or spindle-shaped cells measuring 2–5 μm in diameter. The cells contain elongated or 'cigar-shaped buds' on the narrow base which are diagnostic (Fig. 4.31). Sometimes, fragments of hyphae are attached with the yeast cells.

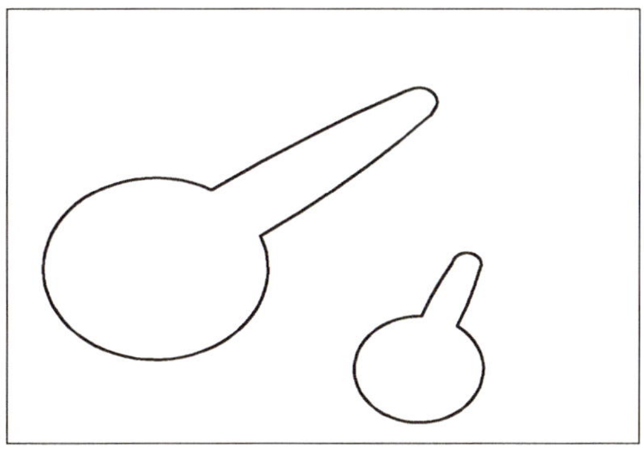

Fig. 4.31 Cigar-shaped bud of *Sporothrix schenckii* yeast cells (Schematic)

Sporothrix schenckii contains an amorphous microfibrillar material in the outermost layer of the cell wall which resembles capsule. Both the mycelial and yeast cells contain a cell wall beneath the capsular material and surrounding the plasma membrane. The cell wall is composed of alkali-soluble and alkali-insoluble glucans. The alkali-soluble glucans of yeasts contain β (1, 3), β (1, 6) and β (1, 4) linkages in a proportion of 44 %, 28 % and 28 %, respectively, whereas the proportion of same linkages differs in alkali-insoluble glucan [66 % β (1, 3), 29 % β (1, 6) and 5 % β (1, 4)]. The β-glucan composition of cell wall remains the same in the mycelial phase. In addition, the yeast cell wall contains a peptide rhamnomannan complex composed of 50 % D mannose, 33 % L rhamnose, 1 % galactose and about 16 % peptides, whereas the mycelial cell wall contains large amount of lipids and protein and a lower quantity of mannose. Melanin and other proteins are also an integral component of yeast cell wall which is involved in fungal adhesion and pathogenesis.

The plasma membrane is present beneath the cell wall in yeast, mycelia and also in conidia. The structure of plasma membrane varies between these three morphological forms. In yeast, the plasma membrane contains scarce but longer invaginations, whereas in mycelia, the invaginations are absent and in conidia, they are abundant but short. There is a vesicle-like structure present in the cell wall which helps in transfer of periplasmic molecules and pigments from the plasma membrane to the extracellular space.

The mould-to-yeast phase transition requires higher temperature (35–37 °C) (except *S. globosa* which grows well in low temperature) and nutrient-rich culture medium. The transition occurs by formation of buds at the tips and along the hyphae. After septation of the hyphae, oidial cell formation takes place. There is no direct budding of yeast from the conidia. The reverse transition (yeast to mould phase) is mostly regulated by calcium which induces RNA and protein synthesis in the yeast cells. It is followed by nuclear division and germ tube formation. The germ tubes become elongated to produce hyphae, and a septum is produced to separate them from the mother cells.

4.14.2 Classification

Sporothrix belongs to the division Ascomycota, class Pyrenomycetes, order Ophiostomatales and family Ophiostomataceae. Previously, *Sporothrix schenckii* was considered as sole species under the genus. Recent study showed existence of five new species under *Sporothrix* species complex on the basis of similarities in macroscopic characteristics, sucrose and raffinose assimilation, ability to grow at 37 °C and the nucleotide sequence (calmodulin). The six sibling species under the complex are *S. schenckii*, *S. albicans*, *S. globosa*, *S. brasiliensis*, *S. mexicana* and *S. luriei*. Other non-pathogenic species include *S. cyanescens* (isolated from the human blood and skin), *S. stylites*, *S. humicola* and *S. lignivora*.

4.14.3 Reproduction

Sporothrix does not undergo any sexual reproduction. Although the evidences suggest that it produces recombination in the nature, some studies related with 18S rDNA analysis identified *Ophiostoma stenoceras* as the sexual form of *S. schenckii*. Later, due to differences in morphological, physiological and virulence properties, it was concluded that *Ophiostoma stenoceras* and *S. schenckii* are two different species.

4.14.4 Susceptibility to Disinfectants

S. schenckii is sensitive to 70 % ethanol, sodium hypochlorite (500–1,000 ppm), hydrogen peroxide (6,000 ppm) and formaldehyde.

4.14.5 Natural Habitat and Distribution

Sporothrix is commonly found as a saprophyte on living and decaying vegetation, woods, leaves and branches of the plants, animal excreta and soil. Sphagnum moss, rose thorns (infection in human is sometimes known as rose gardeners' disease) and hay are the most common plants which can harbour the fungi. The organic materials present in the soil such as cellulose, pH range between 3.5 and 9.4, temperature of 20–25 °C and humidity (more than 90 %) are required for the survival of *Sporothrix* in nature. The humidity requirement varies between the species of *Sporothrix*. Earlier it was thought that armadillo (*Dasypus septemcinctus*) is a potent reservoir of *Sporothrix* as the armadillo hunters often suffer from sporotrichosis. However, it was detected that armadillos do not harbour the fungi in their intestine or skin, whereas the grasses used by the armadillos to make the nests are the major source of the fungi. Other living creatures related to *Sporothrix* transmission are rodents, cats, dogs, squirrels, horses, birds especially parrots, insects and reptiles.

Sporothrix causes the most widespread subcutaneous mycosis throughout the world. The endemic zones are found in India, Japan, Peru, Mexico, Colombia, Uruguay, Guatemala, Brazil and African countries. The largest human outbreak was detected in South Africa which was originated from wooden beam used in the mines. Cases are also reported from United States, France, Spain, Italy, China and Thailand.

In India, West Bengal and North Eastern states such as Assam and Manipur are considered as endemic zone of Sporotrichosis. Sporadic cases are reported from Himachal Pradesh, Sikkim, Tripura, Meghalaya and Nagaland.

4.14.6 Genome

S. schenckii possesses six to eight chromosomes of 460–6,200 kb, with a total genome size of approximately 28 Mbp. The average guanine cytosine (GC) content of the genome is 54.7 mol%. The genome is diploid containing approximately50fg DNA per cell in both the filamentous and yeast phases. The mitochondrial DNA (mtDNA) analysis helps to establish the fungal types in epidemiological study. So far, 32 mtDNA types of *S. schenckii* are detected which is based on restriction enzyme analysis of mtDNA with HaeIII.

4.14.7 Isolation, Growth and Colony Characteristics

S. schenckii can be isolated in Sabouraud dextrose agar with chloramphenicol and in the media containing cycloheximide (mycobiotic agar). The plates are incubated at room temperature (25–28 °C) for 2–7 days. The primary mould colonies are creamy, wrinkled and leathery which become dark coloured at the centre with advancement of age.

The dimorphic conversion into the yeast phase can be detected in enriched media such as blood agar, chocolate agar and brain–heart infusion agar. The conversion requires incubation at 37 °C for 3–5 days with 5–7 % CO_2 tension. The yeast colonies are creamy and yellow to tan in colour.

The production of conidia is better detected in potato dextrose agar and corn meal agar.

4.14.8 Antigenic Characteristics

The major antigen of *S. schenckii* is cell wall glycopeptides (peptidorhamnomannan). The antigenicity varies with the rhamnose/mannose molar ratio depending on the cultural conditions. The peptide portion is composed of threonine, serine, aspartic acid and glutamic acid.

The peptidorhamnomannans can react with the carbohydrate-binding protein concanavalin A, but the rhamnomannan is nonreactive. The reactive fraction is extracted and is known as *S. schenckii* conA-binding fraction (SsCBF). It is used in serodiagnosis for the detection of IgG in human and feline patients. The SsCBF is composed of three glycoproteins, i.e. gp70, gp84 and gp58. Among them, gp70 is the major antigen which is present in both yeast and mycelial phases. The protein (gp70) acts as adhesin to bind with the extracellular matrix protein such as fibronectin and mice dermis.

4.14.9 Virulence Factors

The virulence factors explored by different investigations till date are enlisted in Table 4.34.

4.14.10 Transmission

The transmission of *S. schenckii* occurs through the traumatic inoculation of the fungi into the skin and subcutaneous fat associated with scratches, small injuries and pricks with thorns of plants (roses). Certain job-related and free-time activities, such as gardening, fishing, hunting, farming and mining, are associated with the transmission. Human-to-human spread and occurrence of epidemics are rare.

The zoonotic transmission takes place following the bite or scratches by animals such as cats, dogs, squirrels, armadillos and insects [fire ants (*Solenopsis*)]. The largest outbreak of human

Table 4.34 Virulence mechanisms and factors possessed by *S. schenckii*

Virulence factors	Functions
Thermotolerance	The strains of *S. schenckii* which can tolerate and grow at 37 °C are able to produce lymphatic sporotrichosis, whereas the strains devoid of thermotolerance property are capable of producing cutaneous lesions only
Melanin	Both the yeast and mycelial phases can synthesise melanin. The conidia utilise 1,8-dihydroxynaphthalene (DHN) pentaketide pathway, whereas 3,4-dihydroxy-L-phenylalanine (L-DOPA) is utilised by hyphae for synthesis of melanin. Following roles of melanin in pathogenesis are established
	(a) Melanin produced in the conidia can resist the host cell phagocytosis and diminishes the respiratory burst mediated by human monocytes and murine macrophages. The melanin acts as scavenger for free radicals
	(b) The melanin-producing strains are more invasive in experimental animals than the albino strains
	(c) The melanin pigment may hamper the treatment of sporotrichosis
Fibronectin, laminin and type II collagen-binding proteins, gp70	They act as adhesin, and they are required for attachment of the yeast or mycelium with the host cells and dissemination of infection throughout the body. Laminin-binding protein is expressed by both the yeast and mycelial phases, whereas fibronectin binding protein is expressed by yeast cells only and it is associated with virulence. The gp70 protein prefers to bind with the dermis
Capsular material	The capsular material helps in adhesion of fungi with the host cells
Ergosterol peroxide	It is synthesised by yeast cells to protect themselves from reactive oxygen species which is generated by the phagocytes for killing of the yeast cells
Peptidorhamnomannans	It is present in the cell wall, and it causes suppression of immune response until the sixth week of infection and may act as adhesin to bind the fungi with the host cells
Acid phosphatases	This enzyme is produced by both the yeast and mycelial phases and the conidia of *S. schenckii*. The enzyme helps in intracellular survival of the fungi after phagocytosis
Proteinase I and II	They can hydrolyse the stratum corneum cells in vitro

sporotrichosis due to cat bite or scratches occurred in Brazil (1998–2001), in which 156 cases (out of 178) had direct contact with the cats infected with *S. schenckii*. The cats are considered as the most potent animals capable of zoonotic transmission probably due to the production of large amount of yeast cells. The yeast cells can directly infect the human even in the absence of cat scratches or biting history.

Rarely, *S. schenckii* is transmitted through inhalation route producing extracutaneous form of sporotrichosis, i.e. granulomatous pneumonitis. The laboratory-acquired infection is also reported during culturing of the fungi without proper precautions.

4.14.11 Pathogenesis

Clinical manifestation varies with the load and pathogenicity of the fungi along with the host immune status. After transmission, the mycelial phase is converted into the yeast cells due to exposure to the higher temperature or unidentified signal. The yeast cells are either localised in the subcutaneous tissues or spread via lymphatics or blood producing either fixed or lymphocutaneous or disseminated clinical forms, respectively. The disseminated form occurs either due to immunosuppression or due to inoculation of huge quantity of yeast cells from the infected cats at the multiple locations of the host.

The fixed form is characterised by ulcerated lesion with erythematous edges. Most of the clinical cases (75 %) manifest the lymphocutaneous form which starts with the formation of a papule at the site of trauma. The papule is converted into a subcutaneous nodule with ulceration and pus formation. Secondary lesions arise along the course of lymphatics. However, the involvement of lymph node and other vital organs is rare in human. The disseminated form

Table 4.35 Major diseases of animal and human caused by *S. schenckii*

Fungi	Host	Disease
Sporothrix schenckii	Dogs	The dogs are most commonly affected among the animals. Lymphocutaneous form is most predominant. Other forms such as cutaneous and disseminated sporotrichosis are also detected. The primary lesion develops as subcutaneous nodule (1–3 cm in diameter) at the site of trauma which ulcerates and discharges pus. The dissemination may occur along the course of lymphatics which produces cording and new nodules. Occasional involvement of lymph node is noticed. The lesions are neither painful nor producing itching
	Cats	The primary lesions are small, draining, puncturing type of wounds which appear in the head, base of the tail or the legs (Fig. 4.33). With the progression of the infection, the lesions ulcerate and become nodular. Occurrence of multiple skin lesions is more than the lymphocutaneous form. Systemic involvement is common and is not associated with immunosuppression. The systemic involvement may occur in the bone, lung, liver, spleen, testes, gastrointestinal tract and central nervous system
	Horses	Cutaneous nodules develop along the lymphatics usually on the legs. Systemic involvement may occur like other animals such as cats
	Donkey, mule, pigs, fowl, goat, cattle	Naturally occurring sporotrichosis is detected
	Human	It is also known as 'Rose gardeners' disease' and is characterised by the formation of subcutaneous nodule. Four clinical forms are detected, i.e. fixed cutaneous, lymphocutaneous, disseminated and extra cutaneous. In adults and children, the lesions appear in arm and face, respectively. Lymphocutaneous form is the most predominant

shows multiple skin lesions at the noncontiguous sites, and it involves the bones, joints, lungs, meninges and osteoarticular system. The inhalation of spores rarely produces the extracutaneous form which is characterised by haematogenous dissemination and formation of osteoarticular lesion in most of the cases.

In cats, occurrence of multiple skin lesions is more (39 %) than the lymphocutaneous form (19 %). Systemic involvement is common and is not associated with immunosuppression. The frequency of granuloma formation is low (12 %), and huge amount of fungal elements are detected in the skin which make them more susceptible to *S. schenckii*.

4.14.12 Disease Produced

The major animal and human diseases produced by *S. schenckii* are enlisted in Table 4.35.

4.14.13 Immunity

The innante immune response of the host largely depends on the interaction between the host pattern recognition receptors (PRR) and the fungal pathogen-associated molecular pattern (PAMP). This interaction is not very much explored in *S. schenckii*. The experiments suggest the possible interaction between the host TLR4 and fungal cell wall lipid which can activate the polymorphonuclear cells to phagocytose the yeast cells of *S. schenckii*. The fungi also activate the alternative complement pathway. The complement product (C3b) deposits on the fungal cell wall and acts as opsonin to help in the phagocytosis. After phagocytosis, the superoxide anions, produced by the macrophages, can act as fungistatic or fungicidal.

The cell wall components of *S. schenckii* induce granuloma formation in the host which is a T cell-dependent inflammatory and protective response. The CD4+ T cells produce IFNγ which stimulates the macrophages. The pro-inflammatory cytokines such as IL1 and TNFα are produced by the peritoneal exudate cells (PEC) after 8–10 weeks of the infection. The TNFα further acts on activated macrophages to release nitric oxide. The nitric oxide is the key lethal factor for the yeast and conidia of *S. schenckii*. The nitric oxide also acts as a mediator which can prime the macrophages for potent oxidative burst. The Th1-mediated immunity thus acts as key factor to prevent the dissemination of infection. In contrast, the Th2 response participates in the advanced stages of sporotrichosis only.

4.14.14 Diagnosis

4.14.14.1 Clinical Specimens

The clinical specimens for sporotrichosis include pus or exudates from the deeper part of the lesion, saliva, urine, blood, cerebrospinal fluid, synovial fluid and tissue biopsy specimens.

4.14.14.2 Laboratory Examination

1. *Direct examination*: The wet mounts with 10 % KOH are employed on the clinical specimen for detection of the yeast cells which is diagnostic of sporotrichosis. However, detection of yeasts depends on the fungal burden in the body. So detection is easier in cat specimens than human clinical samples due to high fungal burden in cats. Staining can be performed with Gram stain or Giemsa stain (for diluted pus sample). They are not so much confirmatory because 'cigar-shaped buds' are present only in 1–2 % cases. Instead, fluorescent staining offers confirmatory diagnosis of the yeasts.
2. *Isolation and identification*: Media and incubation condition as described earlier will serve the purpose.
3. *Histopathology*: The yeast cells of *S. schenckii* can be demonstrated in the tissues with haematoxylin and eosin (H&E), Gomori methenamine silver (GMS) or periodic acid–Schiff (PAS) stain. The cells appear as spherical with PAS-positive capsule.
The granuloma produced by the fungi consists of three zones. The central or chronic zone contains neutrophils, histiocyte and lymphocytes along with 'asteroid bodies'. Surrounding the central zone, there is a second zone which contains epithelioid cells and giant cells (strange body and Langerhans type). The third and outer zone is composed of lymphocytes, plasmocytes and fibroblasts. However, the zones are not a constant diagnostic feature and the cells can mingle with each other. In dogs, the granuloma is characterised by increased neutrophil infiltration and absence of lymphocytes and macrophages which can differentiate it with the leishmaniasis lesion.
The 'asteroid bodies' or 'asteroid corpuscles' (found in 20–66 % cases) originate from the Splendore–Hoeppli reaction in sporotrichosis. The Splendore–Hoeppli reaction is a localised immunological response in fungal, bacterial or parasitic infection, and it is characterised by the formation of radiating homogenous, refractile, eosinophilic club-like material surrounding a central eosinophilic focus (Fig. 4.32). The asteroid bodies in Sporotrichosis consists the viable yeast cells in the

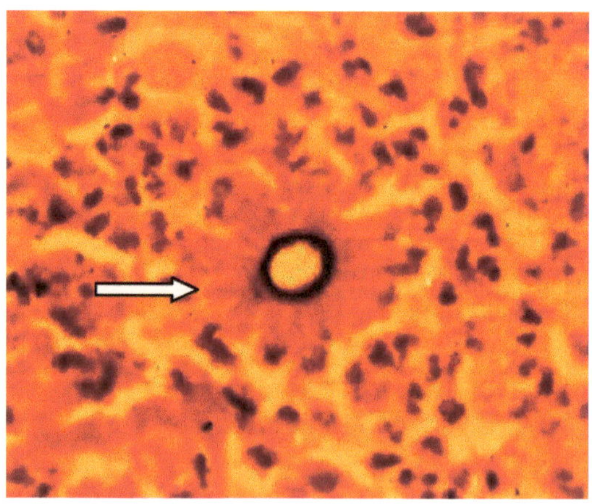

Fig. 4.32 Splendore–Hoeppli reaction in sporotrichosis. *Arrow* indicates the radiating club-like material surrounding a central eosinophilic focus (Photograph courtesy: Prof. P.P. Gupta, Ex-Director of Veterinary Research, Punjab Agricultural University, Punjab, India)

centre along with the host IgG and IgM on the spikes of the radiated crown. Thus, the asteroid bodies use the host immune molecules to provide a safe shelter to the yeasts.

4. *Serological tests*: The serological tests such as agarose gel precipitation test (AGPT), tube agglutination (96 % sensitivity), latex agglutination (94 % sensitivity) and immunoelectrophoresis can be employed for detection of antibodies against *S. schenckii*. The AGPT does not cross-react with the sera from leishmaniasis. However, these tests are not so much sensitive for the diagnosis of cutaneous form of the infection. Enzymatic immune assay such as ELISA is a preferred choice of serodiagnosis. The test was initially developed against the *S. schenckii* conA-binding fraction (SsCBF) antigen. However, in serodiagnosis of cutaneous form, this antigen-based ELISA produces cross-reaction with other infection such as paracoccidioidomycosis. Later, the exoantigens produced by a mycelial phase of *S. schenckii* strain, isolated from Brazil, were used in ELISA which produced no cross-reaction with antigens and serum samples from patients with coccidioidomycosis, histoplasmosis or paracoccidioidomycosis.
5. *Skin test for detection of delayed-type hypersensitivity (DTH)*: The 'sporotrichin test' is a tuberculin-like skin test for the detection of DTH against sporotrichosis. The antigen used for the test varies with the country and accordingly the result also varies. The 5 McFarland standard suspension of heat-killed yeast cells, extracted mycelial antigens at 1:2,000 dilution (approximately 5×10^7 cells/mL) or diluted yeast phase antigens at 1:4,000 were used as antigen in Brazil and Mexico, respectively. The antigen is inoculated at 0.1 mL into the forearm or back of the suspected patient. A positive result involves an indurated, erythematous and painful area of more than 5 mm in diameter at the site of the injection 48 h after inoculation. The test is sensitive for detection of lymphocutaneous and fixed cutaneous forms. Rarely, the false-negative reaction is produced in anergic or immunosuppressed patients and false-positive reaction is produced in healthy patients previously suffered from the infection. Experimentally the sporotrichin test with SsCBF antigen in guinea pig was found reactive. In Brazil, the sporotrichin test was carried out among the captive mammals present in a zoo for epidemiological investigation.
6. *Molecular biology*: The chitin synthase gene (ChS1)-based PCR was first developed which can detect up to 10 picogram of genomic *S. schenckii* DNA. Further PCR-based assays are also developed, detecting DNA topoisomerase II and internal transcriber space in the rRNA gene of *S. schenckii*.

4.14.15 Treatment

Potassium iodide (KI) or sodium iodide (NaI) is commonly used for the treatment of sporotrichosis especially for the lymphocutaneous and fixed forms of the infection in cats, dogs and horses. The compounds have high efficacy with minimal side effects. It is easy to administer and is relatively low in cost in comparison to the standard antifungals. KI has immunostimulation property which can activate the neutrophils and monocytes to destroy the fungi. Some authors claim its proteolytic nature which helps to resolve the granulomas. However, exact mechanism of action is not known. The KI therapy should be continued for 2–3 months for complete cure. However, in human, some side effects such as nausea, vomition, diarrhoea, abdominal pain and metallic taste are noted. Some cats are also susceptible to iodide toxicosis.

In human, itraconazole is the first choice treatment with the dose rate of 100–200 mg/day orally (for lymphocutaneous and fixed forms). It has minimal toxicity and good tolerance even in long-term treatment except during pregnancy. In cats and dogs, also itraconazole is preferred nowadays. For disseminated form of sporotrichosis, amphotericin B is preferred even in pregnant women (after 12 weeks of pregnancy) and immunosuppressed patients. Fluconazole is less effective and is preferred only for those patients

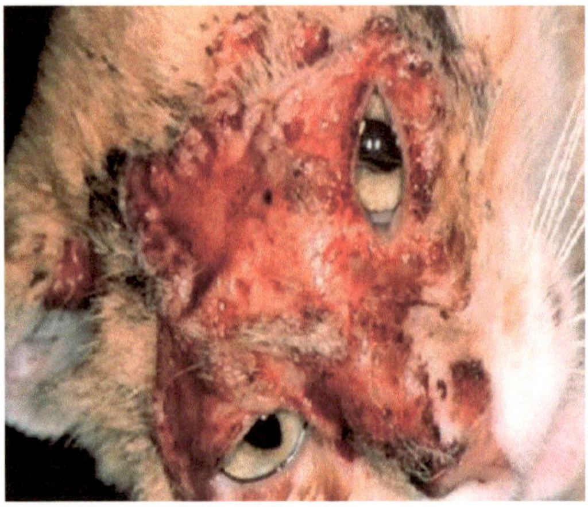

Fig. 4.33 Sporotrichosis in a cat (Photograph courtesy – Prof. Alexandro Bonifaz, Head, Department of Mycology and Dermatology service, General Hospital of Mexico, Mexico City)

who are allergic to itraconazole. The ketoconazole is not used due to poor efficacy and high level of toxicity. The terbinafine and posaconazole showed good in vitro activity against *S. schenckii*. The thermotherapy is recommended for fixed cutaneous form in pregnant or drug-allergic patients because *S. schenckii* cannot grow at more than 40 °C. The hot baths (45 °C) for 15–20 min three times a day and topical thermal compression are advised (Fig. 4.33).

4.15 Mycetoma (*Madurella*, *Pseudallescheria*, *Scedosporium*)

In ancient Sanskrit writing (*Atharva Veda*), the first description of mycetoma (*Pada Valmikan*) was documented. In the seventeenth century, a German physician (Engelbert Kaempfer) working in India first reported clinical human cases of mycetoma which is followed by case reports of French missionaries in Pondicherri, India (eighteenth century). In 1842, John Gill, a British Army physician working at South India, also reported a case of human mycetoma. Godfrey (1846) first described an authentic clinical report of four human mycetoma cases from Madras (Chennai), India, and named them as 'morbus tuberculosis pedis'. Due to its widespread occurrence among barefooted natives in Madurai (India), it was popularly known as 'Madura foot'. In 1860, Carter established the term 'mycetoma' and its etiological correlation with the fungi. In 1874, Carter wrote a monograph entitled 'On mycetoma or the fungus disease of India' where he established firmly the correlation between the fungus and the menace.

Laveran (1902) first isolated *Streptothrix mycetomi* as the fungal agent from the mycetoma (black grain). Brumpt (1905) reclassified *Streptothrix mycetomatis* as *Madurella mycetomi*. British Medical Research Council (1977) further changed the name from *Madurella mycetomi* to *Madurella mycetomatis* which is accepted worldwide.

Radaeli (1911) first isolated a fungus (*Monosporium apiospermum*) from the white-grain mycetoma in Sardinia. Currently this fungus is known as *Pseudallescheria boydii* (teleomorph).

Scedosporium as a fungal causative agent of human otitis was first described in 1889. Saccardo (1911) first illustrated its etiological role in human mycetoma and suggested the name as *Scedosporium*.

4.15.1 Aetiology

Mycetoma (maduromycosis) is chronic pyogranulomatous infection of the skin and subcutaneous tissues caused by either fungi (eumycetoma) or bacteria (actinomycetoma). A wide variety of fungi such as *Madurella mycetomatis*, *Pseudallescheria boydii* (previously *Allescheria boydii*), *Trematosphaeria grisea* (previously *Madurella grisea*), *Aspergillus*, *Fusarium*, *Curvularia*, *Acremonium*, *Paecilomyces* and others cause mycetoma in human and animals (Table 4.36). Among them, *Madurella mycetomatis* is the major causative agent of black-grain mycetoma in human which is usually found in tropical region (India, Sudan and Madagascar). In temperate and subtropical region (United States, Brazil), *Pseudallescheria*

Table 4.36 Fungal species associated with mycetoma and properties of grains produced

Fungi	Grain colour	Size and texture
Madurella mycetomatis	Black	0.5–4 mm, hard
Pseudallescheria boydii	White	0.5–1 mm, soft
Trematosphaeria grisea (*Madurella grisea*)	Black	0.3–0.6 mm, hard
Aspergillus flavus	Green	0.5–3 mm, soft
Aspergillus nidulans	White	0.5–4 mm, soft
Fusarium oxysoprum	White	0.5–1 mm, soft
Curvularia geniculata	Black	0.5–1 mm, hard
Acremonium falciforme	White	0.5–1 mm, soft
Falciformispora senegalensis (*Leptosphaeria senegalensis*)	Black	0.5–2 mm, hard
Neotestudina rosati	White	0.5–1 mm, hard
Exophiala jeanselmei	Black	0.2–0.3 mm, soft
Cylindrocarpon cyanescens	White	–
Polycytella hominis	White	–
Plenodomus avramii	Black	–
Corynespora cassiicola	Black	–
Pyrenochaeta mackinnonii	Black	–
Phialophora verrucosa	Black	–

boydii is the most common etiological agent of eumycetoma in human.

In addition, *Microsporum canis*, *Cladophialophora bantiana* and *Curvularia lunata* are also detected as aetiology of mycetoma in cat, dog, and dog and parrot, respectively.

4.15.2 Morphology

The mycetomal lesion consists of hyphal aggregates, known as 'grain'. The morphology and colour of grain help in the identification of fungal species (Table 4.36). Most of the grains are round or oval and trilobed. Two types of grains are detected, i.e. filamentous and vesicular (rare). The filamentous grain is composed of branched, septate and brown hyphae. The vesicular grains consist vesicles and light-coloured hyphae at the centre and cementing substance at the periphery (melanin, heavy metal, protein and lipid). The cementing substances interfere the penetration of host immune effector molecules and antifungals.

Microscopic study revealed that the *thallus* of *Madurella mycetomatis* is composed of regularly septate hyphae. Most of the cultures are sterile. Sometimes conidiation is observed which produces two types of conidia. Round conidia are detected in flask-shaped *phialides*, and pyriform *conidia* are observed in branched or unbranched conidiophores without phialides. In *Trematosphaeria grisea* (*Madurella grisea*), two types of hyphae are observed, i.e. thin (1–3 μm width) and broad (3–5 μm width).

Pseudallescheria boydii (teleomorph) has more than one anamorph, i.e. *Scedosporium apiospermum* (anamorph) and *Graphium eumorphum* (synanamorph). *P. boydii* is a homothallic fungus. The reproductive structure (*cleistothecium*/*ascocarp*) is produced in nutritionally poor media. It initiates with the formation of coiled ascogonia which is converted into matured cleistothecia (100–300 μm) within 10 days. The cleistothecial wall is composed of thin, flat, polygonal brown cells, and it contains globose or subglobose ascus which bears eight ascospores inside. The ascospores (4–5 μm × 7–9 μm) are unicellular, oval or ellipsoidal and yellow-brown. The ascospores contain an oil droplet inside (Fig. 4.34).

Scedosporium (anamorph) contains hyaline, septate and cylindrical hyphae (2–4 μm in diameter). The conidiogenous cells emerge from the hyphae which produce oval, brown, sticky conidia (4–9 μm × 6–10 μm) which do not

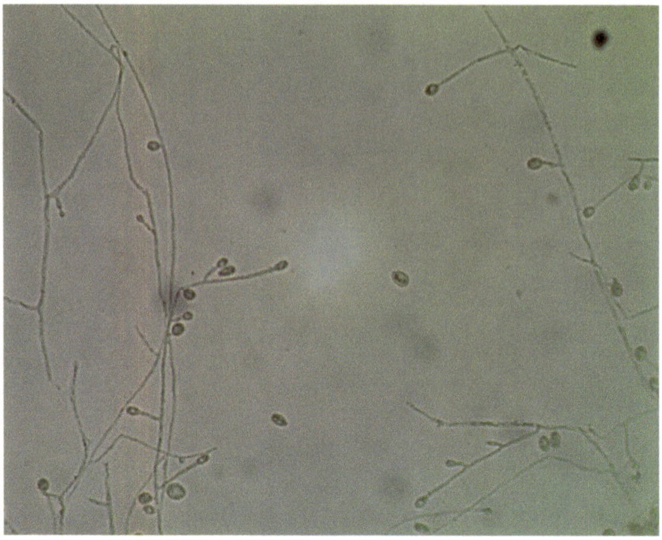

Fig. 4.34 Cleistothecium and ascospores of *Pseudallescheria boydii* (Photograph courtesy: Professor Ali Zarei Mahmoudabadi, Jundishapur University of Medical Sciences, Ahvaz, Iran)

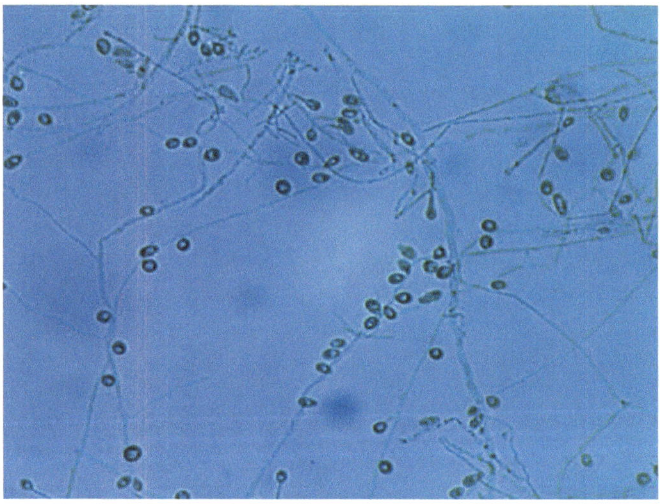

Fig. 4.35 Conidial arrangement of *Scedosporium* (Photograph courtesy: Professor Ali Zarei Mahmoudabadi, Jundishapur University of Medical Sciences, Ahvaz, Iran)

contain any oil droplet inside. These conidia are produced as single from one conidiophore (solitary annelloconidia) and are truncated at the base (Fig. 4.35).

The *Graphium* synanamorph mycelium is composed of erect, stiff, olive-brown bundles of hyphae, terminating in a brush of slender conidiogenous cells. The conidiophores are cemented together to form synnemata (united conidiophores) to produce clusters of conidia. The conidia are extra slender and less pigmented than *Scedosporium* conidia, however, truncated at the base-like *Scedosporium* conidia (Fig. 4.36).

4.15.3 Classification

Classification of *Madurella* was possible with the help of molecular tools (sequencing of internal-transcribed spacer (ITS) and beta-tubulin genes) because the genus lacks sexual reproduction or specialised morphological features and detection of asexual conidia is also extremely difficult. Based on nuclear sequence data, *Madurella* is placed under two different orders, i.e. Sordariales and Pleosporales. The type species *M. mycetomatis* belongs to the order Sordariales along with *M. pseudomycetomatis*, *M. fahalii* and *M. tropicana*, whereas

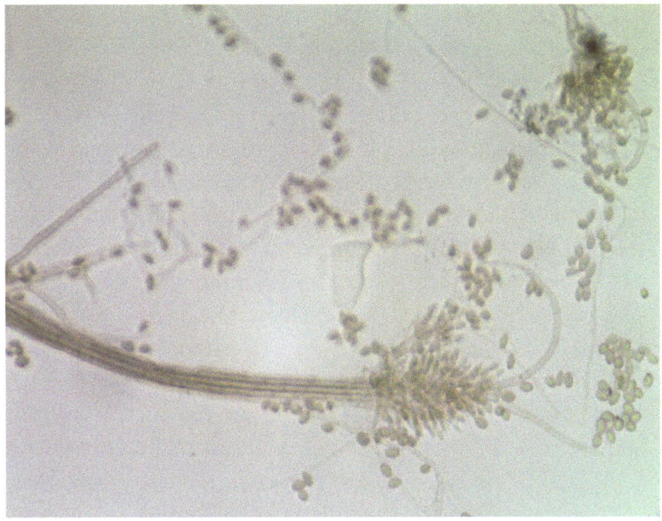

Fig. 4.36 Synnemata and conidia of *Graphium* (Photograph courtesy: Professor Ali Zarei Mahmoudabadi, Jundishapur University of Medical Sciences, Ahvaz, Iran)

Trematosphaeria grisea (previously *Madurella grisea*) belongs to the order Pleosporales.

Pseudallescheria boydii (teleomorph) and *Scedosporium apiospermum* (anamorph) belonged to phylum Ascomycota, class Euascomycetes, order Microascales and family Microascaceae. The genus *Pseudallescheria* is composed of seven species such as *P. boydii*, *P. africana*, *P. angusta*, *P. desertorum*, *P. ellipsoidea*, *P. fimeti* and *P. fusoidea*. Among them, *P. boydii* is the only pathogenic species. Molecular phylogenetic analysis revealed that *P. boydii* is a species complex consisting 44 haplotypes.

4.15.4 Susceptibility to Disinfectants

M. mycetomatis is susceptible to 70 % ethanol and moist heat (60 °C for 30 min).

4.15.5 Natural Habitat and Distribution

Madurella resides in the soil and sometimes in thorns of the plants (*Acacia*) which also helps in the transmission of the infection. *Pseudallescheria boydii/S. apiospermum* is detected in brackish water, saltwater, sewage, soil, swamps, coastal tidelands, manure of poultry, cattle and bat guano. They can tolerate high temperature, low-oxygen tension and high salt concentration (5 %). So they are common in polluted environmental niches having poor air circulation and high osmotic pressure.

Mycetoma may occur throughout the world but is common in tropical and subtropical zone (between the latitudes of 15S and 30N) with major foci in Asia (southern India), Africa (Somalia, Senegal, Sudan, Mauritania, Niger, Ethiopia, Chad, Cameron) and South and Central America (Mexico, Argentina, Brazil, Venezuela) ('mycetoma belt'). Sudan has the highest incidence of mycetoma in human which is followed by Mexico. The endemic zone should have arid and hot climate with limited rainfall. The cases are also reported from the United States (Texas, California), Germany, Egypt, Turkey, Philippines, Japan, Lebanon, Thailand, Iran, The Netherlands, Ceylon and Saudi Arabia.

4.15.6 Genome

The mitochondrial genome of *M. mycetomatis* is a compact, circular DNA molecule (45,590 bp). Most part of genome is AT rich and the overall GC content is only 26.8 %. The genome encodes for the rRNAs, tRNAs (27), proteins of respiratory chain complexes (11), ATP synthase subunits and 6 intronic proteins including the ribosomal protein (rps3).

The genomic organisation of *Pseudallescheria boydii* is still unknown.

4.15.7 Isolation, Growth and Colony Characteristics

The grains are collected from the mycetomal lesion for isolation of etiological fungi or bacteria. The grains are washed with sterile saline and antibiotics, crushed with sterile glass rod and are spread over the plate. The medium commonly used is Sabouraud dextrose agar with antibiotics such as gentamicin sulphate (400 mcg/mL), penicillin G (20 U/mL), streptomycin (40 mcg/mL) and chloramphenicol (50mcg/mL) for isolation of *Madurella*. Some species of *Madurella* are inhibited by antibiotics, so one plate without antibiotic should also be kept. Recently, the use of trypticase soy agar with 5 % horse serum has been proposed to enhance the growth of *M. mycetomatis*. *M. mycetomatis* grows at 37 °C, while *M. grisea* grows at 30 °C.

Sabouraud glucose agar is preferred for isolation of *P.boydii/Scedosporium apiospermum* which requires 25 °C as optimum growth temperature. This fungus can tolerate up to 37–42 °C, low-oxygen tension or anaerobic condition, high concentration of magnesium chloride (5 %) or sodium chloride (5 %). Addition of benomyl, amphotericin B or dichloran makes the medium selective for isolation of *P. boydii/ Scedosporium apiospermum.*

The colonies of *M. mycetomatis* are white, flat, woolly or leathery at the beginning which later produce a diffusible brownish pigment and become folded and heaped due to production of aerial mycelium. The sporulation is better observed in straw extract agar, wheat extract agar, soil extract agar and water agar. The colonies of *M. grisea* are leathery, folded with radial grooves, light brown to greyish in colour which later become dark brown to reddish-brown and have a brownish-black reverse.

The colonies of *P. boydii* in Sabouraud glucose agar look different from the upper surface (obverse) and from the reverse. The colour is initially white and later becomes dark grey or smoky brown (obverse), while from the reverse, the colonies appear pale with brownish-black zones. During preservation of the culture for many years, the colonies become dirty white coloured and have fur-like surface without any conidia.

4.15.8 Antigenic Characteristics

M. mycetomatis secretes enzymes such as fructose-bisphosphate aldolase (FBA) and pyruvate kinase (PK) which act as immunodominant antigens. The antibodies against them are detected in the sera of patients.

P. boydii has a peptidopolysaccharide antigen (peptidorhamnomannan) which has a similarity with *Sporothrix schenckii* peptidorhamnomannan, but they do not cross-react during serological reactions.

4.15.9 Virulence Factors

The virulence factors of *M. mycetomatis* and *P. boydii* are enlisted in Table 4.37.

Table 4.37 Virulence mechanisms and factors possessed by *M. mycetomatis* and *P. boydii*

Virulence mechanisms/ factors	Functions
Ability to grow at 37 °C	It helps in better adaptation of *M. mycetomatis* to the host body temperature
Grain formation	Within the host tissues, both *M. mycetomatis* and *P. boydii* produce grain (sclerotia or granules) to evade the host immune response. Grains are fungal aggregates along with a matrix or cementing substance
Peptidase (28 KDa, 34 KDa)	The extracellular peptidases are released by *P. boydii*. The peptidases split host fibronectin and laminin which may help to escape the fungi from host effector cells such as fibronectin-activated macrophage and monocyte
Serine protease (33 KDa)	It is released by *S. apiospermum* which splits host fibrinogen and helps in tissue invasion

4.15.10 Transmission

In the endemic zone, *M. mycetomatis*, *Scedosporium* and other fungal and bacterial agents can grow saprobically in the environment (thorns, metal piece, wire, wood splinter, nails, speculated seeds, etc.). During penetrating trauma, puncture or lacerated wound of the host, these etiological agents can enter the subcutaneous tissue through the contaminated objects. Sometimes the objects (thorn) are detected within the mycetoma lesion. There is no evidence of transmission from animal to human or human to human.

In human, there is no racial or ethnic predominance. However, males aged between 16 and 30 years are more susceptible than females. In a study in West Bengal (India), occurrence of mycetoma between male–female was 183: 81.

4.15.11 Pathogenesis

Following entry of the organism through the wound, the fungi produce an aggregate with a cementing substance for their survival. This fungal aggregate is known as 'grain' (sclerotia or granules). Sometimes the grains do not possess any cementing substances, and the hyphal cell walls become thickened near the periphery of the grain which can protect the fungi from host immune system. The host innate immune system attempts to clear the infection through activation of complement cascade and complement product-mediated chemotaxis of granulocytes towards the fungal grains. In natural infection, granulomas develop which contain the fungal grains at the centre, surrounded by a zone of polymorphonuclear neutrophils (PMN). Sometimes the PMNs are detected within the grains also. With the progress of the infection, the granulomas mature and contain lymphocytes, macrophages, plasma cells, giant cells and mononuclear cells.

4.15.12 Disease Characteristics

The clinical features of eumycetoma in human and animals are described in Table 4.38.

Table 4.38 Clinical features of mycetoma in human and animals

Species	Clinical features
Human	The incubation period is unknown. Mean duration of eumycetoma is 9.8 years. The infection starts with a small, firm, painless tumour-like mass in subcutaneous tissue which further spreads into the subcutaneous fat, muscle, ligaments and bones. With the progress, multiple nodules develop, ulcerate and drain through the sinuses. The sinus tracts appear within 3 months–1 year. The discharge is purulent or serosanguinous and it contains the grains. The draining sinuses may heal and a new one opens simultaneously. The lesion is painless due to nerve damage by fibrous tissue reaction. Intense pain is felt only when there is bone involvement and secondary bacterial infection. Most of the mycetoma cases (70 %) occur in the feet. It may occur at other sites such as the hand, knee, arm, leg, head, neck and face, thigh, perineum and rarely at chest and abdominal wall, eyelid, testis, vulva, middle ear, paranasal sinus, mandible and lymph nodes
Horses	Mycetoma in horses is characterised by the subcutaneous nodules which may suppurate and drain a serosanguineous or purulent discharge through multiple sinus tracts. The discharges contain the fungal grains
Cats	Subcutaneous nodules are detected in dorsal trunk which drains purulent exudate containing yellow-coloured granules
Dogs	Nodules are observed in subcutaneous tissues and sometimes in the ribs and thorax with visible granules. Rarely disseminated mycetoma is found which involves the lymph nodes, heart, kidneys, adrenal glands, bone, pancreas, brain and eyes
Goats	Mycetomal lesions are observed in hind legs and scapula
Cattle	Lesions are found in the skin, nasal cavity and lymph node
Grand Eclectus parrot (*Eclectus roratus roratus*)	Black nodules were detected in thoracic cavity and lung causing central nervous system disorder and wing paralysis

4.15.13 Immunity

Macrophages are the first line of defence against *Scedosporium apiospermum* which are able to phagocytose both the hyphae and conidia. The hyphae are opsonised with the serum, and the macrophages can produce oxidative burst/superoxide anion (O_2^-) to control the opsonised hyphae. The polymorphonuclear neutrophils (PMN) can also participate in phagocytosis and oxidative burst-mediated killing of *Scedosporium*. However, the cytokine (IL15) can enhance this oxidative burst of PMNs against *Scedosporium prolificans* but not against *Scedosporium apiospermum*.

The cell wall α-glucan of *P. boydii* is recognised by toll-like receptor (TLR2) and CD14 of host macrophages and dendritic cells. The interaction can transduce the signal to secrete pro-inflammatory cytokines.

4.15.14 Diagnosis

4.15.14.1 Clinical Specimens

The diagnosis of mycetoma depends on three major clinical symptoms (triad of clinical symptoms) such as edematous subcutaneous tissue, multiple draining sinuses and excreation of grains. The clinical specimens include pus, exudate, bandage gauze and biopsy tissues which are used for naked eye examination to detect the grains. The physical properties of the grains are indicative of the etiological agent (Table 4.36).

Fine needle technique is also used for the collocation of grains from the lesions. A fine needle attached to a syringe is inserted within the lesion, and it is moved up and down in 3–4 different locations to collect the exudates. The bloody material is kept for some time to clot.

4.15.14.2 Laboratory Examination

1. *Direct examination*: For the detection of etiological agent (fungi or bacteria), the grains are crushed and fixed with 10 % formal saline on a microscopic slide. The slide can be stained with modified acid-fast staining to detect the bacterial etiological agent, if any. For the detection of fungal aetiology, the slides are immersed in 20 % KOH followed by fluorescence microscopy using the Uvitec, Blankophor or calcofluor white. However, the fluorescence microscopy cannot differentiate between the fungal genera.
2. *Isolation and identification*: Media and incubation condition as described earlier will serve the purpose.
3. *Histopathology*: The grains can be identified in haematoxylin and eosin (H&E)-stained tissue sections. Gram–Weigert stain can be used to detect bacterial mycetoma and periodic acid–Schiff (PAS) and Gomori methenamine silver, and Gridley stains are used to detect hyphae and fungal spores in eumycetoma. The grains have differentiating characteristics in the tissue section which creates an idea regarding the aetiology. For example, the black grains of *Trematosphaeria grisea* (*M. grisea*) have a non-pigmented centre and dark-coloured periphery with brown cementing substances. Further, the tissue reaction is also suggestive for the fungal aetiology. The grains are surrounded by neutrophils in *M. mycetomatis* infections and by eosinophils in *P. boydii* infections. However, most of the fungi-causing eumycetoma cannot be differentiated by histopathology. For example, the fungal species such as *Scedosporium*, *Aspergillus* and *Fusarium* produce hyaline hyphae with septation at regular intervals and dichotomous branching. Sometimes *Scedosporium* produce terminal or intercalary chlamydospores which are confused with the yeasts.
4. *Serological tests*: Indirect haemagglutination assay (IHA) and immunodiffusion test (ID) are developed for the detection of *Pseudallescheria boydii*. For the detection of *M. mycetomatis* antibodies in human, immunodiffusion and counter immune electrophoresis (CIE) are widely used. Sometimes, ID tests produce false-negative reaction especially during the early phase of the infection and when the treatment is ongoing. The CIE is found to be superior than ID for better

sensitivity. ELISA is also used for the detection of antibodies. Previously the ultrasonicated mycelial extract (cytoplasmic protein) was used as antigen in serological tests which produce variable results and cross-reaction with other fungi. The use of recombinant antigens [fructose-bisphosphate aldolase (FBA) and pyruvate kinase (PK)] in ELISA produces promising result for detection of *M. mycetomatis* antibodies.

5. *Molecular biology*: The PCR is developed for the detection of *M. mycetomatis* and *Pseudallescheria boydii* using the primer for internal-transcribed spacer (ITS) located between 18S and 28S genes and beta-tubulin (*BT2*) genes, respectively. Rolling circle amplification (RCA) technique is used to detect *P. boydii* which is based on the amplification of the ITS region with pan-fungal primers. The loop-mediated isothermal amplification (LAMP) method is also developed for the confirmation of *P. boydii* which do not require sophisticated instruments and is suitable for endemic zones of developing countries.

4.15.15 Treatment

In human, ketoconazole or itraconazole with surgical excision is recommended for eumycetoma. Further, treatment with cotrimoxazole-streptomycin, trimethoprim and sulphametoxazole, and tetracyclins or rifampicine also produces promising result. The study with experimental *Scedosporium* infection in laboratory mice revealed that posaconazole, fluconazole and voriconazole are effective in reduction of fungal burden.

In animals especially in horses, surgical excision with thiabendazole powder applied topically is found effective against *M. mycetomatis* infection.

4.16 *Pythium*

Smith (1884), a veterinarian working in India, first described a chronic cutaneous granulomatous disease in horses known as 'bursattee' in local language. Later the aetiology of the disease was correlated with the fungi based on histology. In 1901, Dutch investigators (de Haan and Hoogkamer) working in Indonesia isolated the fungi and identified them as Zygomycetes (due to lack of sporulation) from the horses having similar kind of infection (Hyphomycosis destruens equi). Witkamp (1924), another Dutch scientist, also isolated the fungi from the horses showing a similar kind of syndrome. In 1974, Austwick and Copland recognised them as a member of the genus *Pythium* under oomycete family. Based on this finding, Chandler et al. (1980) proposed the term pythiosis for the infection. In 1987, de Cock first identified them as a new species under the genus *Pythium* and proposed the nomenclature of *Pythium insidiosum*.

4.16.1 Morphology

Pythium can produce wide (4–10 μm), hyaline, coenocytic or sparsely septate hyphae which produce branches at right angle (lateral pegs). In older cultures, sporangia-like swellings (12–30 μm) are observed. However, it is not considered as a true fungus due to lack of chitin and sterol in cell wall and cytoplasmic membrane, respectively. The cell wall is composed of cellulose and β-glucans.

4.16.2 Life Cycle

The motile zoospores are considered as an infectious form of *P. insidiosum*. They are released in the aquatic environment and infect the damaged skin or gastrointestinal mucosa of animals and human. The formation and release of zoospores take place from the zoosporangium. The zoosporangia are produced from the hyphae which are in contact with certain plants. The long, narrow and thin-walled tubular structure with widen tip is developed from zoosporangia (discharge tube). The sporangial protoplasm enters the discharge tube to produce a terminal vesicle. The vesicle expands in size and it is released from the discharge tube as a zoospore. The zoospores are motile with two flagella present in anterior (tinsel) and posterior (whiplash) site. The tinsel flagellum produces the thrust and the whiplash flagellum acts as rudder during the swimming of the zoospores.

The swimming of the zoospores is directed by chemotaxis (calcium, sugars and amino acids released from the damaged plant root, animal skin or tissues), electrotaxis and auto-aggregation. The aggregation of zoospores in a site attracts more zoospores to adhere which is known as auto-aggregation. The zoospores adhere with the damaged skin surface of the animals and human by formation of a cyst-like structure. The gelatinous sticky glycoprotein covering of the cyst helps in adherence. Germ tubes are produced from the cyst which generates hyphae. The hyphae penetrate the host tissues to establish the infection.

4.16.3 Classification

Pythium belonged to the kingdom Stramenopila, class Oomycetes, order Pythiales and family Pythiaceae. Currently there are 250 numbers of species under the genus *Pythium*. Most of them are nonpathogenic, even some are beneficial except the *P. insidiosum* which causes diseases in mammals. Molecular phylogenetic studies also revealed its distant relationship with the fungal kingdom. All the *Pythium* species are divided into 11 different phylogenetic groups (A–K) and *P. insidiosum* belonged to group C. Further molecular studies based on heterogeneity of rRNA genes separated *P. insidiosum* into three different clades based on geographical locations. The clade I isolates belonged to North, Central and South America, clade II isolates are from Asia and Australia, whereas clade III isolates are from Thailand and the United State. The clade I and II isolates are closely related, but clade III is significantly different from the others.

Recently similarity between one of the phylogenetic clade (K) of *Pythium* and *Phytophthora* was detected, and the new genus *Phytopythium* has been proposed. *Pythium* is necrotroph and zoospores are produced from a vesicle in the zoosporangia, whereas *Phytophthora* is hemibiotroph and the zoospores are produced directly within the sporangia.

4.16.4 Reproduction

Some strains of *P. insidiosum* can undergo sexual reproduction to produce oospores. The generation of oospores occurs in the oogonium (female structure) which is intercalary, smooth and subglobose, whereas the antheridium (male structure) is diclinous (unisexual). The antheridium produces a rigid fertilisation tube from their tip during conjugation. The oospores are aplerotic which cannot fill the oogonium completely and often pressed to one side of the oogonium.

The asexual reproduction generates zoospores which are described earlier.

4.16.5 Susceptibility to Disinfectants

Pythium are susceptible to 0.5 % solution of sodium hypochlorite, quaternary ammonium compounds.

4.16.6 Natural Habitat and Distribution

P. insidiosum resides in the freshwater sources. Other species under the genus *Pythium* may inhabit in the soil, marine water (*P. grandiosporangium*) or freshwater. Some species are parasitic to insect (mosquito), algae, fish and other fungi. They are used as a biological control measure against the mosquitoes such as *Aedes aegypti*, *Culex quinquefasciatus* and treehole mosquito (*Ochlerotatus sierrensis*).

Pythiosis is common in animals and human of Southeast Asia including India, Indonesia, Japan, Korea, New Guinea and Thailand; coastal Australia and New Zealand; and South America such as Argentina, Brazil, Colombia, Venezuela,

Costa Rica, Guatemala, Haiti, Panama and Nicaragua, North America, Mexico and the United States.

4.16.7 Genome

The pyrosequencing study of plant pathogenic *Pythium* genome (*P. aphanidermatum*, *P. arrhenomanes*, *P. irregulare*, *P. iwayamai*, *P. ultimum var. sporangiiferum*, *P. vexans*, *P. ultimum var. ultimum*) revealed the smaller genome size than other members of the family sequenced so far (*Hyaloperonospora*, *Phytophthora*). The annotation study showed the presence of 11,957–15,291 numbers of predicted genes in the *Pythium* genome. *Pythium* species have 1.5–1.7 introns per gene and average exon length is 312–630 bp. All the sequenced genome has common 76 clusters containing 782 secreted proteins (core *Pythium* secretome).

4.16.8 Isolation, Growth and Colony Characteristics

P. insidiosum can be isolated in common fungal media such as Sabouraud dextrose agar, vegetable extract agar, peptone yeast glucose agar, cornmeal agar, potato flakes agar and V8 agar with streptomycin (200 mcg/mL) and ampicillin (100 mcg/mL). As an alternative, Campy blood agar with trimethoprim, vancomycin, polymyxin B, cephalothin and amphotericin B can be used for isolation. The small pieces of fresh, non-macerated tissues are directly placed on the surface of plates. The plates can be incubated both at 25 or 37 °C along with a beaker filled with distilled water to provide the moisture.

The typical colonies are observed within 12 –24 h. The colonies are submerged, white to colourless, and have an irregular radiate pattern. In the laboratory, zoospore production can be induced by keeping boiled grass blades on the surface of a 1–2-day-old colony growing on 2 % water agar and incubating at 37 °C for 18–24 h.

4.16.9 Antigenic Characteristics

The antigens of *P. insidiosum* can induce humoral immunity in the host, although it cannot clear the infection effectively. The immunodominant antigens of *P. insidiosum* are 28 KDa, 30 and 32 KDa (twin antigen), 34 KDa, 38 KDa and 74 KDa proteins. Among them, recent study indicated the 74KDa protein as major immunodominant antigen. Further, there are two diffuse proteins (49 KDa, 51 KDa) detected only in the animal strains of *P. insidiosum* but not in human strains.

4.16.10 Virulence Factors

The virulence factors of *P. insidiosum* are enlisted in Table 4.39.

4.16.11 Transmission

Direct contact with *P. insidiosum* zoospores-infested water is the major route of transmission in animals and human. The damaged skin or gastrointestinal mucosa helps in the adherence of the zoospores and dissemination of infection. In horses, most of the pythiosis lesions occur in the legs and ventral abdomen (Fig. 4.37). Recently, *P. insidiosum* is also isolated from the larvae of *Culex quinquefasciatus* (tropical mosquito) in India which indicates the possible role of mosquito in propagation of infection.

The large, male, young (1–3 years) and outdoor working dog breeds (e.g. Labrador

Table 4.39 Virulence mechanisms and factors possessed by *P. insidiosum*

Virulence factors	Functions
Protease	*P. insidiosum* produces three or more proteases of different molecular weight. They are considered as serine proteases. The proteases help in host tissue invasion
Mechanical force	The mechanical force is exerted by the tips of the elongating hyphae which also help in host tissue invasion

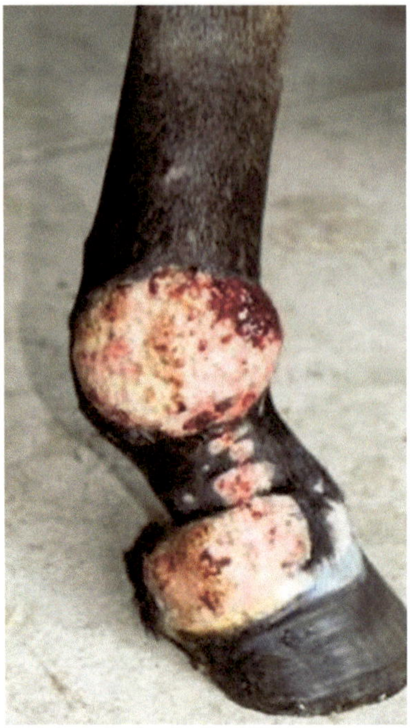

Fig. 4.37 Ulcerative lesion of pythiosis in a horse (Photograph courtesy: Prof. P.P. Gupta, Ex-Director of Veterinary Research, Punjab Agricultural University, Punjab, India)

retrievers) are more susceptible to the infection. In human, agriculture or water leisure activities are the major predisposing factors. However, no zoonotic transmission is detected so far.

4.16.12 Pathogenesis

After adherence of the *P. insidiosum* zoospores with the damaged skin or gastrointestinal mucosa, the cysts are produced. The encysted zoospores produce a germ tube stimulated with the host body temperature. The germ tube develops into hypha which invades the host tissues with the help of the virulence factors (Table 4.39). In human, the hyphae also penetrate the blood vessels and disseminate throughout the body. Sometimes the blood vessel penetration causes thrombosis. The infection is life-threatening in human and animals if not treated early.

4.16.13 Disease Produced

The major human and animal diseases produced by *P. insidiosum* are enlisted in Table 4.40.

4.16.14 Immunity

The antigens of *P. insidiosum* can induce humoural immunity (Th2 mediated) in the host which generates precipitate and agglutinate anti-*P. insidiosum* antibodies. These antibodies are nonprotective and can be only used for the diagnosis of the infection, whereas the cell-mediated immunity (Th1 mediated) produced by activated macrophages, mast cells, eosinophils and other inflammatory cells causes extensive tissue damage, but it can effectively clear the infection. During natural infection, *P. insidiosum* always presents the antigens in such a manner that the antigen-presenting cells (APC) secrete IL4 which converts the Th0 cells into Th2 cells and produces Th2-mediated immunity. In vaccinated or cured patients, higher amounts of IFNγ and IL2 are detected, indicating the generation of Th1 response which has played a protective role.

4.16.15 Diagnosis

4.16.15.1 Clinical Specimens

The 'kunkers' from the horses can be collected and transported into the laboratory in saline water with antibiotics (streptomycin and ampicillin, tetracycline, chloramphenicol). Other specimens include skin scrapings, tissue biopsies (deep wedge biopsies), exudates and fine needle aspirates from lymph node. Transport of the clinical samples in ice is not recommended because the low temperature (4 °C) can inhibit the growth of *P. insidiosum*.

4.16.15.2 Laboratory Examination

1. *Direct examination*: The wet mount with 10 % KOH can be prepared directly from the clinical specimens. The presence of

Table 4.40 Major diseases of animal and human caused by *P. insidiosum*

Host	Disease
Horse	Cutaneous and intestinal forms are detected. In cutaneous form, large, rounded, granulomatous, nodular and ulcerative lesions are observed, composed of eosinophils and the fungal hyphae. The tissues contain 'kunkers' (necrotic yellow-grey materials) which is a specific symptom of pythiosis in horses. The kunkers are produced due to degranulation of eosinophils by the invading hyphae. New eosinophils replace the older ones which accumulate in the tissues as a necrotic mass. The clinical signs associated with cutaneous form are itching, biting of the wound area and auto-mutilation, lameness, enlargement of regional lymph nodes, anaemia and hypoproteinaemia. The cutaneous form is known as 'swamp cancer' or 'Florida horse leeches'. In intestinal form (rare), stenotic fibrous and GI tract lesions are observed
Dogs	Intestinal form is more frequent than the cutaneous form. The symptoms are vomition, weight loss and intermittent diarrhoea. There is segmental transmural thickening of the stomach, small intestine, colon, rectum, oesophagus or pharynx. Enlargement of mesenteric lymph node is observed which becomes palpable in the mid-abdomen. In cutaneous form, the lesions are found in the face, leg, ventrum, perineum and tail
Cattle	Pythiosis in cattle is a sporadic infection which is common in subtropical area during rainy season. Granulomatous lesions are observed in the limbs. The animals cannot stand and in untreated cases death may occur due to dehydration. Secondary bacterial infection especially with anaerobes may occur producing 'infectious pododermatitis'
Sheep	Granulomatous lesions are observed in limbs and rhino-pharyngeal area. Clinical symptoms include nasal discharge, swelling of nostrils and face
Cats	Pythiosis is rare in occurrence. Both cutaneous and intestinal forms are reported. The cutaneous lesions are detected in legs, digits or footpads
Camel, Bear, Jaguar, Bengal Tiger, Californian nestling white-faced ibis bird	The pythiosis lesions are observed in face and vulva of camel, preputial gland and gastrointestinal tract of bear, gastrointestinal tract of tiger, and wings, neck, head and limbs of Californian nestling white-faced ibis birds
Human	The infection is noticed in apparently healthy persons having contact with the contaminated water. There are four clinical forms of human pythiosis which are cutaneous (infecting the face or limbs as a granulomatous and ulcerating lesion), vascular (affects arteries and causes arterial occlusion and aneurysm), ocular and orbital (corneal ulcers) and disseminated pythiosis (vital organs)

coenocytic or sparsely septate hyphae in the slide indicates pythiosis.

2. *Isolation and identification*: Media and incubation condition as described earlier will serve the purpose. The ideal clinical specimens for culture are kunkers, tissue biopsies. The samples are washed thrice with the sterile saline solution before culture and are cut into 5–10 mm blocks. The tissue blocks are impregnated directly into the agar plates.
3. *Histopathology*: The histological sections are stained with H&E, Gomori methenamine sliver (GMS) and periodic acid–Schiff (PAS). However, the hyphae are clearly visible only with GMS stain. Affected tissues contain multiple foci of necrosis surrounded by neutrophils, eosinophils and macrophages. The granulomas are composed of macrophages, plasma cells, multinucleate giant cells and fewer neutrophils and eosinophils. The fungi are found within necrotic area or at the centre of granulomas.
4. *Serological tests*: In the laboratory, different serological tests such as immunodiffusion, ELISA, immunochromatographic assay and Western blot are developed for the detection of anti-*P. insidiosum* antibodies. Immunodiffusion test is however unable to detect the antibodies in dogs, whereas ELISA shows potential in early detection of postoperative recurrence. A simple and reliable haemagglutination test is also developed in which agglutination of sheep red blood cells coated with a *P. insidiosum* extract is possible using the serum samples of the patients.

The immunohistochemical techniques include immunofluorescence and immunoperoxidase staining which can be used in paraffin-embedded fixed tissues.

5. *Molecular biology*: The *P. insidiosum*-specific PCR is developed using the internal-transcribed spacer (ITS) of rRNA gene as a target. The PCR is successfully applicable for the confirmation of DNA extracted from the culture, fixed tissues and clinical specimens if preserved properly either at −70 °C or at ambient temperature with 95 % ethanol. Recently a nested PCR is also successfully employed for the detection of *P. insidiosum* from paraffin-embedded tissues.

4.16.16 Treatment

Surgical intervention is the choice of treatment in animals and human. In cutaneous pythiosis, amputation of legs is recommended when the lesions appear in the extremities. However, chances of recurrence are also higher (45 %) after surgical removal of cutaneous lesions in horses and dogs. In animals with gastrointestinal pythiosis, segmental lesions should be resected with 3–4 cm margins. Due to anatomical location, certain area such as oesophagus, gastric outflow tract, rectum and mesenteric root cannot be removed completely.

Due to the absence of ergosterol in the cytoplasmic membrane, *P. insidiosum* is resistant against most of the antifungals such as ergosterol synthesis inhibitors (azole group). However, the azole group of antifungals (itraconazole, ketoconazole, miconazole and fluconazole), terbinafine and amphotericin B can also alter the permeability of fungal cytoplasmic membrane and thus may have little antifungal effect against *Pythium*. So, the antifungal therapy with itraconazole (10 mg/kg by oral route in every 24 h) and terbinafine (5–10 mg/kg by oral route in every 24 h) is recommended for 2–3 months after surgery to reduce the chances of recurrence.

Currently combination of azole group and caspofungin (inhibitor of β-glucan synthesis) produces promising result against *P. insidiosum*. Immunotherapy with killed ultrasonicated mycelium or secreted antigens from broth culture after precipitation produced rapid cure of the lesions in horses and dogs, when combined with surgery or antifungals.

4.17 Emerging and Uncommon Pathogenic Fungi

4.17.1 Phaeohyphomycosis

Phaeohyphomycosis is used to describe the fungal menace ranging from cutaneous colonisation to abscess formation and dissemination, caused by a group of melanised fungi which reproduce in culture by unicellular budding. The fungi belong to phylum Ascomycota, subphylum Pezizomycotina and class Dothideomycetes. There are two orders, i.e. Capnodiales and Pleosporales, under the class Dothideomycetes. The order Capnodiales consists of *Cladosporium cladosporioides* and the order Pleosporales consists of *Alternaria alternata*, *A. infectoria* and *Bipolaris spicifera* as major aetiological agent. The other aetiological fungi such as *Cladophialophora bantiana*, *Exophiala attenuata*, *Fonsecaea multimorphosa* and *Phialophora verrucosa* belong to the order Chaetothyriales under the class Eurotiomycetes. The poultry black yeast (*Ochroconis gallopava*) belongs to order Ochroconiales under the class Eurotiomycetes.

The black fungi mycelia are composed of cylindrical hyphae with thick cell wall in which melanin is deposited. Some of them (*Exophiala*) also have yeast-like phases and they are known as 'black yeast'. The conidiogenous cells emerge from the hyphae and produce conidia. The method of conidiogenesis (blastic or thallic) and physical properties of conidia help in the identification of fungal species (Table 4.41; Figs. 4.38, 4.39, 4.40, and 4.41).

The melanised fungi (*Alternaria*, *Bipolaris*) are ubiquitous saprobe inhabiting soil and plants. The black yeasts (Chaetothyriales) can survive in polluted environment also rich in aromatic hydrocarbons. Their ability to use hydrocarbons

Table 4.41 Physical properties of conidia and colony characteristics produced by different melanised fungi

Fungi	Conidia	Colony characteristics
Cladosporium	The conidiophores are branched or unbranched. During conidiogenesis, ramoconidia (shield cells) are produced first which give rise to branching chains of dark-coloured, single- or double-celled conidia with prominent attachment scar (hila)	Colonies are olivaceous to black and velvety
Alternaria alternata	Chains of conidia are found which are large, dark, euseptate and muriform	Colonies are woolly, pale to olivaceous to black. The growth of the colony is rapid
A. infectoria	The conidia have long apical beaks which serve as secondary conidiogenous cells	
Bipolaris spicifera	There is bipolar germination of conidia. The conidiophores are geniculate (bent) with flattened hilum. *B. spicifera* has three distosepta (pseudosepta where inner walls are involved) and four cells	Colonies are woolly, grey to black, with rapid growth
Exophiala	The conidiogenous cells produce conidia by short percurrent proliferations (annellations). The tip of an annellide increases in length and becomes narrower as each subsequent conidium is formed. Sometimes phialides are also present	Colonies are primarily mucoid which later become more filamentous
Fonsecaea	The conidia are produced from swollen denticles (small projection). The secondary and tertiary conidia in chains (four conidia) are also produced on *sympodial* conidiophores or on phialides	Colonies are olivaceous to black and velvety
Phialophora verrucosa	It has dark, funnel-shaped *collarettes*	Colonies are olivaceous to black and velvety
Ochroconis gallopava	The conidia are clavate and borne from denticles	Colonies are brown, velvety with a red diffusing pigment

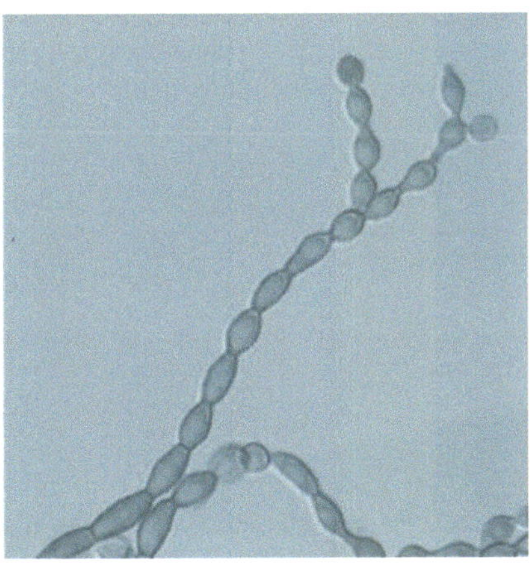

Fig. 4.38 Conidial arrangement of *Fonsecaea multimorphosa* (Photograph courtesy: Prof. Sybren de Hoog, CBS Fungal Biodiversity Centre, The Netherlands)

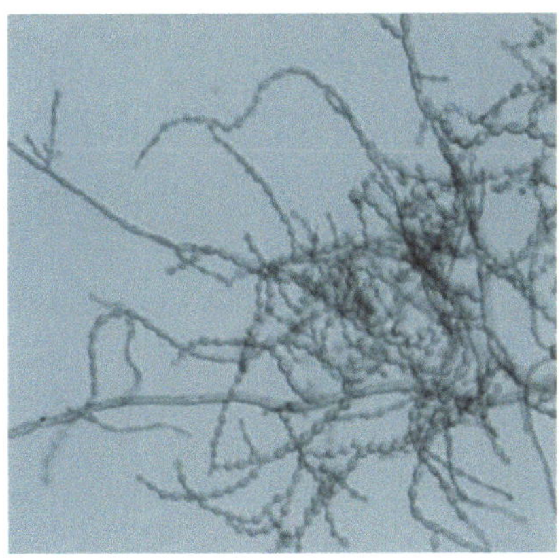

Fig. 4.39 Conidial arrangement of *Fonsecaea multimorphosa* (Photograph courtesy: Prof. Sybren de Hoog, CBS Fungal Biodiversity Centre, The Netherlands)

explains about their affinity for central nervous system due to the structural similarity between the hydrocarbon and neurotransmitter.

Most of the species are able to grow at 37 °C or at higher temperature (*Cladophialophora bantiana*, *Ochroconis gallopava*) in Sabouraud dextrose agar, brain–heart infusion agar, potato dextrose agar, malt extract agar, V8 juice agar, cereal agar, carnation leaf agar, cornmeal dextrose agar (nonselective) or cycloheximide-containing

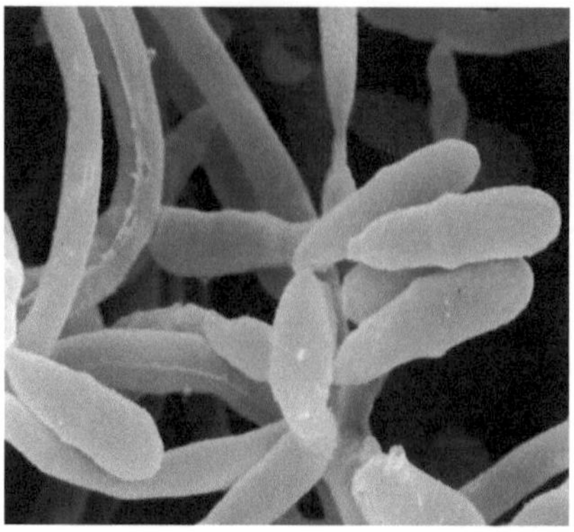

Fig. 4.40 Conidia of *Ochroconis gallopava* (Photograph courtesy: Prof. Sybren de Hoog, CBS Fungal Biodiversity Centre, The Netherlands)

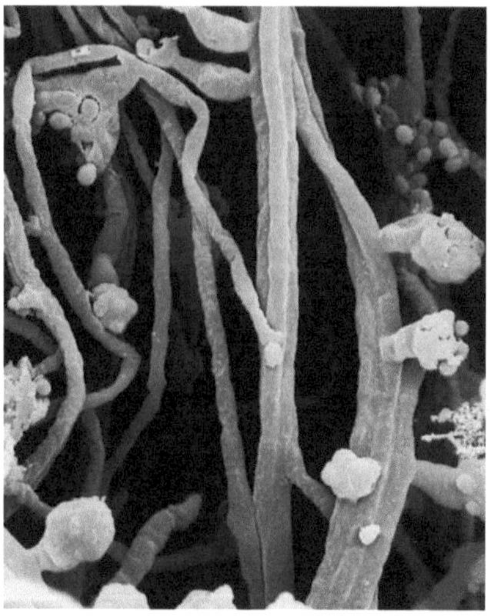

Fig. 4.41 *Phialophora verrucosa* (Photograph courtesy: Prof. Sybren de Hoog, CBS Fungal Biodiversity Centre, The Netherlands)

Table 4.42 Virulence mechanisms and factors possessed by melanised fungi

Virulence factors	Functions
Melanin	The fungi-obtained melanin from dihydroxynaphthalene, a cell wall constituent. The pigment prevents the lytic effects of hydrolytic enzymes on the fungal cell wall and scavenging free radicals released by phagocytic cells during the oxidative burst. Thus, melanin helps in evasion of immune response. It also provides resistance against physicochemical agents, toxic metals, desiccation, ionising radiation and antifungals
Cell wall	The cell wall is unusually thick where the melanin can be deposited
Yeast-like phase	It helps in evasion from phagocytosis
Aspartic protease (*Alternaria*)	It degranulates eosinophils
Additional conidiogenesis	Transmission of infection
Thermotolerance (*Ochroconis gallopava*, *Exophiala*), osmotolerance, adhesion, hydrophobicity, production of extracellular polysaccharides, siderophores and acidic or alkaline secondary metabolites	These fungal properties help in the establishment of infection
Assimilation of aromatic hydrocarbons	It helps in the survival of the fungi in polluted environment

media (act as selective media except for *Ochroconis gallopava*). Antibiotics are also added to suppress the bacterial contamination. All of the fungal species produce velvety or woolly colonies with pigmentation in the media (Table 4.41).

There are several virulence factors and mechanisms used by the melanised fungi to establish an infection (Table 4.42).

The transmission of *Ochroconis* in poultry occurs through the inhalation of fungal elements from the contaminated litter. In sheep, *Cladosporium cladosporioides* is transmitted through inhalation in a humid and overcrowded pen. Minor skin abrasions and trauma also play a role in the transmission of melanised fungi in human and animals.

The Chaetothyriales group of black yeasts prefer to infect invertebrates and cold-blooded vertebrates with a moist skin and having water-associated life. Black yeast infection of mammals is unusual due to the presence of water repellent skin and the capacity to produce granuloma with lymphocytes, granulocytes and giant cells (highly developed immune system). In debilitated mammals such as immunocompromised animals or wild animals kept in captivity, chances of infection are more.

Ochroconis gallopava was reported to cause epidemic encephalitis in poultry, turkey, captive owl and cats. It causes cerebral phaeohyphomycosis in quail chicks and grey-winged trumpeters (*Psophia crepitans*). The birds showed sudden ataxia, intermittent torticollis and rigidity of legs.

In cats, cutaneous, subcutaneous and systemic phaeohyphomycoses were reported which were caused by *Cladophialophora bantiana*, *Phialophora verrucosa*, *Fonsecaea pedrosoi*, *Exophiala spinifera*, *Alternaria alternate* and *A. infectoria*. The infection showed nasal, ocular, renal and cerebellar involvement.

In dogs, *Ochroconis gallopava* and *Bipolaris spicifera* were detected to cause systemic phaeohyphomycosis which involved the brain, bone and kidney. *Cladosporium cladosporioides* was detected to cause granulomatous encephalitis and nephritis. Onychorrhexis (breakage of nails) on the digits and blackened corium, black plaque-like lesion in mouth was observed in dogs suffering from *Alternaria* infection.

In sheep (merino breed), *Cladosporium cladosporioides* causes systemic phaeohyphomycoses which was characterised by anorexia, fever and respiratory distress followed by death. In horse, firm brown-coloured skin nodule, alopecia and scaling in head and neck were detected due to *Alternaria* infection.

The clinical specimens include tissue biopsies, aspirates and body fluids. The postmortem samples such as the brain sections can also be collected from the animals died due to suspected encephalitis. The blood from the suspected animals is collected for culture if the lesions are not accessible. Gram stain, KOH wet mount and fluorescent staining (calcofluor white) are the most commonly used methods for the direct examination of specimens. The histopathological staining of brain tissues with haematoxylin and eosin shows the invasion of septate, melanised hyphae. The histological sections of the nodules in cats reveal granulomatous dermatitis containing neutrophils, macrophages and giant cells. The pigmented, yeast-like cells and hyphal elements are identified in the tissue sections of nodules in cutaneous or subcutaneous form and black granular necrotic areas throughout the liver in systemic form. Other histopathological stains such as melanin Fontana–Masson, periodic acid–Schiff (PAS) and Gomori methenamine silver can be used for the detection of light-melanised fungi. Isolation of the melanised fungi in artificial culture is required to confirm the infection. However, the specimens from the patients receiving antifungals may not produce any growth, and in such cases histopathology helps in identification. The nucleic acid sequencing of the internal-transcribed spacer (ITS1 of rRNA gene) is developed for the detection of black yeast (Chaetothyriales) infection in animals.

Treatment of melanised fungi in human and animals is difficult due to the presence of melanin in their cell wall which make them refractory to the most of the antifungals. In animals, the systemic antifungals such as amphotericin B, itraconazole, voriconazole, posaconazole, caspofungin and anidulafungin are used for the treatment of black yeast infection. Clean and hygienic pen and poultry houses can prevent the transmission of black yeast.

4.17.2 *Pneumocystis*

Carlos Chagas (1909) first observed the cystic forms of *Pneumocystis* in the lungs of rats and guinea pigs. Vanek and Jirovec (1952) first described *Pneumocystis* as an aetiological agent of endemic pneumonia in children. Later (1960 onwards) it was recognised as an opportunistic pathogen in immunocompromised children. In 1980s, *Pneumocystis* was recorded as most widespread opportunistic pathogen in individuals suffering from AIDS. With the advancement of antiretroviral therapy, the incidence of *Pneumocystis* infection in human subsided.

Earlier *Pneumocystis* was considered as protozoa. The molecular phylogenetic study based on 16S rRNA and mitochondrial gene established its correlation with the kingdom Fungi. Recent classification included *Pneumocystis* under the phylum Ascomycota, class Archiascomycetes, order Pneumocystidales and family Pneumocystidaceae. *Pneumocystis* infects mammals only and is highly host specific. Previously all the forms of *Pneumocystis* was designated as *Pneumocystis carinii* as an honour to Antonio Carini, an Italian biologist who described them in rats. At present *P. carinii* is used to describe the rat or other animal-related forms of *Pneumocystis.* It is detected in dogs, cats, sheep, goat, monkey, rat and mice. Human form of *Pneumocystis* is designated as *Pneumocystis jirovecii* in honour of Jirovec.

In the life cycle of *P. carinii*, two developmental stages exist, i.e. the mature cyst and the trophozoite. The matured cysts are spherical (5–8 μm in diameter) containing up to eight numbers of intracystic bodies (1.2 μm mean diameter). The cysts are surrounded by two cell layers, i.e. outer layer (15 nm thick) and inner layer (35 nm thick). The cyst protoplasm consists of a matrix and intracystic bodies. The matrix contains mitochondria, ribosome, empty vacuoles and membrane debris. The intracystic body is spherical to oval containing a centrally located nucleus, endoplasmic reticulum, free ribosome, mitochondria and glycogen particles. After release of intracystic bodies, the cyst becomes banana shaped which is known as remnants of the ruptured cysts. The intracystic bodies originate trophozoites.

The trophozoites are 0.3 μm in diameter, vary in shape and produce clusters, and they are associated with pneumocytes in the lung tissue. They are haploid in nature and they can undergo binary fission (endogeny). Later two trophozoites conjugate to produce a diploid cell which undergoes meiosis (three times) to produce a spherical cell (cyst) with eight intracystic bodies. The trophozoites are surrounded with a 20–30 nm thick dense coat which contains small electron translucid areas. The coat helps in anchoring and nutrient uptake. The trophozoite cytoplasm is rich in free ribosome, glycogen particles, mitochondria with cristae and endoplasmic reticulum-like structure. A nucleus is present in the variable position of the cell with a nucleolus either centrally or peripherally. The vacuoles, microtubules and a budding Golgi complex are also sometimes observed. Some trophozoites contain filopodium-like structure which helps in making contact with host cells. Both the developmental stages lack sterol (ergosterol) in their cell wall.

Pneumocystis is ubiquitous in nature which is commonly detected in soil, pond water and indoor or outdoor air. The antibodies are detected in children (7 months–4 years) and adults (both in HIV and non-HIV healthy patients) throughout the world without any clinical syndrome, indicating their exposure to the organism. The fungi cause a major opportunistic infection (pneumonia) throughout the world, and the prevalence rate is rising in developing countries of Asia (Thailand, Vietnam, Cambodia), Africa (Tanzania, Congo, Ivory Coast, Zimbabwe, Kenya, Ethiopia, South Africa) and North and South America (Mexico, Panama, Guatemala, Venezuela, Brazil, Chile, Argentina and Peru). In India, the prevalence rate of pneumonia caused by *Pneumocystis* is low (5–6.1 %).

The genome of *P. carinii* is haploid and consists of 13–15 linear chromosomes (300–700 Kbp each). The genes present in the genome contain 60–65 % AT sequences and introns (less than 50 bp, nine per gene in number).

Pneumocystis isolation in artificial culture is not possible probably due to its stringent requirement which cannot be reproduced outside the host.

The major antigen of *Pneumocystis* is a surface glycoprotein (major surface glycoprotein/glycoprotein A, 90–120 KDa), encoded by *msg* gene. The relatively conserved carboxy-terminal region of the protein (MsgC) is immunodominant in nature, and anti-MsgC antibody titre is often detected in infected patients. The Msg protein helps in interaction with a number of host cell proteins (fibronectin, vitronectin, surfactants) and attachment with the host epithelial cells. *Pneumocystis* can secrete proteases [kexin (Kex1, Prt1)] having similarity with serine proteases (Kex2) produced by *Saccharomyces cerevisiae*. The kexin acts as virulence factor by proteolytic processing and activation of Msg protein.

Transmission of *P. jirovecii* occurs from a common environmental source or through person-to-person contact. The animal-to-human transmission of *Pneumocystis* is not observed. The infection can also reactivate from latent childhood contamination during immunocompromised condition. The childhood contamination may remain latent in the host for a prolonged period due to changing capability of *Pneumocystis* Msg structure to evade immune response. The seroconversion rate of *Pneumocystis* in children was detected as 5 % per month. The children may act as reservoir and transmit the infection to other persons who are healthy or immunocompromised. In healthy persons again it maintains the latency, whereas in immunocompromised patients, it produces fatal pneumonia. In animals, inhalation is the major route of transmission.

After transmission, the organisms proliferate extracellularly in the lung alveoli. The virulence factor (Msg) of *Pneumocystis* interacts with alveolar epithelial cells for attachment. Diffuse alveolar lesions are observed which suggest the involvement of host immune response with T cell-mediated inflammation.

P. jirovecii causes pneumonia (PcP) in immunocompromised human patients due to malignancies, immunosuppressive therapy, congenital immunodeficiency or other infection (HIV) and malnourished children. *P. carinii* can infect Arabian foals with or without severe combined immunodeficiency. The clinical signs in horses include dyspnoea, cough and nasal discharge. The lungs become firm, meaty and mottled. The lung parenchyma resists of being cut and bulges on cut section during in vitro test. Naturally occurring *P. carinii* infection is also reported from piglets, goats, dogs, cats and non-human primates. In goats, *P. carinii* infection was associated with *Mycobacterium avium* subsp. *paratuberculosis*.

The natural antibodies (IgM) generated without any antigenic stimulation as a component of innate immunity offered protection against *Pneumocystis*. The natural antibodies help in recognition of fungi by dendritic cells and enhance dendritic cell maturation and migration into the draining lymph nodes. The natural antibodies also help in the development of anti-*Pneumocystis*-adaptive immunity containing both humoural and cellular responses. The anti-*Pneumocystis* humoural immunity plays a major role in protection. The passive transfer of hyperimmune antiserum or monoclonal antibodies conferred protection in experimentally infected mice. As a component of cellular immunity, optimum level of CD4+ T cells (Th1, Th2, Th17) along with functional co-stimulatory molecules (CD2, CD28) is associated with the control of *Pneumocystis* infection. The level of Th2 was detected to be maximum during experimental infection. The natural IgM is required for the development of Th2 response, Th17 cell differentiation and immunoglobulin class switching, whereas CD8+ T cells offer a partial protection only. Further, γδ T cells may play a role in controlling inflammation and pulmonary injury during diminished αβ-T cell concentration.

The diagnosis depends on direct examination of clinical specimens such as sputum, saliva, bronchoalveolar lavage (BAL) and lung tissues. The commonly used stains are Giemsa, methenamine silver, toluidine blue and Diff-Quik. However, direct detection method especially in histological sections requires expertise. Instead, detection of *Pneumocystis* antigen (β-D glucan) in BAL and antibodies in serum is an alternative diagnostic strategy. The PCR is a better choice for diagnosis targeting the genes encoding Msg antigen or mitochondrial large subunit rRNA (mtLSU) gene. The real-time PCR is also developed for the detection of *Pneumocystis* heat shock protein 70 gene (*hsp70*). Currently attempts are in progress with *cdc2* as target gene to increase the sensitivity and specificity of the molecular assay.

Trimethoprim–sulphamethoxazole is the drug of choice for treatment and prophylaxis of *Pneumocystis*-associated pneumonia in human. Some patients with HIV infection (25 %) are unable to tolerate a full course of trimethoprim–sulphamethoxazole. In such cases, intravenous pentamidine and intravenous clindamycin with oral primaquine or atovaquone are effective alternatives. In animals, echinocandin (prevent synthesis of β-D glucan) was detected to be active against matured cyst but not against trophozoite forms. Corticosteroid as adjunct therapy was successful in curing of the infection. Amphotericin B or triazoles are not very effective due to lack of sterol in the cell wall.

4.17.3 *Prototheca*

Zopf and Kühn (1880) isolated an unknown organism from the slime flux of a linden tree which was later characterised and identified as a fungi due to cultural similarity with yeasts (Krüger 1894). In 1913, similarity of the organisms with algae was established due to its internal sporulation (Chodat 1913). Later it was reclassified as algae under the genus *Prototheca* and family Chlorelaceae (West 1916). Davies et al. (1964) first described human infection due to *Prototheca* which was detected from cutaneous lesion of a farmer at Sierra Leone. However, the etiological correlation between *Prototheca* and bovine mastitis was established earlier (Lerch 1952).

Prototheca is an achlorophyllous mutant of the green algae with nonfunctional chloroplast. It was originated from *Chlorella* during evolution, and later it acquired different cell wall composition, physiology and better ability to survive environmental stress. They are unicellular, oval or spherical (3–30 μm in diameter) and surrounded by bilayered cell wall. The cell wall does not contain glucosamine, muramic acids, chitin and cellulose. The nonpathogenic species can produce capsule.

There are five major species under the genus *Prototheca*. Among them, *P. wickerhamii* and *P. zopfii* are pathogenic, whereas *P. stagnora*, *P. ulmea* and *P. moriformis* are considered as nonpathogenic.

Asexual reproduction takes place in which the cytoplasmic cleavage produces endospores/aplanospores/sporangiospores (2–20 in number). The endospores are generated within a sporangium, a thick-walled, oval or spherical body (10–30 μm in diameter) (Fig. 4.42). The

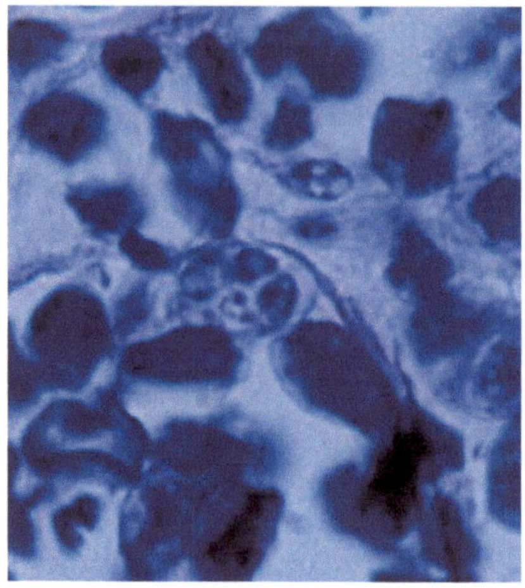

Fig. 4.42 *Prototheca* sporangium with a central rounded endospore (1,000x) in haematoxylin–eosin-stained tissue smear (Photograph courtesy: Prof. Cornelia Lass-Flörl, Medizinische Universität Innsbruck, Austria)

sporangia of *P. wickerhamii* (3–10 μm in diameter) are generally smaller than the sporangia of *P. zopfii* (7–30 μm in diameter). Initially the spores are irregular in shape. The spores increase in size and disrupt the sporangial membrane and thus they are released from the sporangia. The spore release requires 5–6 h time and sufficient nutrients. The newly released spores again increase in size and repeat the cycle.

Prototheca is ubiquitous in nature and widely distributed throughout the world. The organism was isolated from slime flux of trees, soil, sewage, fresh and salt water, urban water supply system, pasture mud, dairy barn drainage and fruits (banana). *Prototheca* prefers to inhabit in the water bodies containing plant-decomposed materials. The organism was also isolated from skin, sputum and faecal samples of healthy human and animals such as cattle, pig, horses and sheep.

The mitochondrial genome of *P. wickerhamii* is 55,328 base pairs in size with more AT content (74.2 %). The genes present in the genome can encode cytochrome oxidase, apocytochrome b, NADH dehydrogenase (*nad1-7*, *nad4L*, *nad9*), ATPase (*atp6*, *atp9*, *atp1*), ribosomal RNAs (5SrRNA, small subunit (srn), large subunit (lrn) RNA), tRNAs and ribosomal proteins. The plant-specific genes such as *orf25*, *orf244*, *orfB* and *rrn5* are detected in the mitochondrial genome of *P. wickerhamii* which indicates their relationship with algae, the simplest plant.

Prototheca is heterotrophic in nature. They require external sources of organic carbon (glucose, fructose and galactose), nitrogen (protein or inorganic nitrogen), oxygen and thiamine for their growth. They can be isolated in blood agar, brain–heart infusion agar and Sabouraud dextrose agar (SDA) without cycloheximide. The specific medium for *Prototheca* should contain folate (for inhibition of bacteria) and 5-fluorocytosine (for inhibition of yeast). The plates are incubated at 25–37 °C for 48–72 h. *P. wickerhamii* colonies are smooth, moist, cream-coloured and yeast-like. *P. zopfii* colonies in SDA are large, dull white, irregularly margined and granular with a central protrusion.

P. zopfii has three biotypes (I, II, III) which are based on carbohydrate assimilation. The biotype II (associated with bovine mastitis) can ferment glucose and glycerol but not galactose (delayed fermentation may occur in some isolates), whereas biotype III (swine isolates) cannot ferment glycerol. The antigenic structure differs between the three biotypes. Based on 18S rRNA gene sequence and cellular fatty acids, the biotype III is recently proposed as a novel species (*P. blaschkeae* sp. nov.). Further, all the three biotypes are currently considered as three genotypes (1–3) of *P. zopfii*.

The transmission of *Prototheca* occurs by direct contact or traumatic inoculation in human, dogs and cats. In cattle, the organism enters from contaminated environment (faeces, wet litter) through direct contact with teat end, especially 25–30 min after milking when the teat canal sphincter is relaxed to prevent the ascending infection. Alternatively, *Prototheca* is also transmitted through intramammary infusion material. The cow-to-cow transmission during milking is also possible. The predisposing factors include poor milking hygiene, prolonged antimicrobial therapy and warm and humid weather. In dogs, transmission may occur via ingestion also which results systemic infection. Man-to-man or man-to-animal and vice versa transmission of *Prototheca* is not detected.

Prototheca (*P. zopfii* genotype 2) causes subclinical or chronic mastitis in cattle without systemic involvement. The clinical syndrome is pain in udder, watery milk secretion with clots and decreased milk production. The somatic cell count varies between 6 and 9 million cells/mL. The infection is limited within the udder and regional lymph node. However, sporangia and sporangiospores are detected within the macrophage and neutrophils in the alveolar lumen and interstitium, indicating the intracellular nature of the infection. So it is difficult to eradicate and it becomes chronic mastitis which persists for several lactation periods. The chronic mastitis becomes progressive interstitial mastitis with consequent alveolar atrophy and severe decreased milk production.

P. zopfii (sometimes *P. wickerhamii*) causes systemic protothecosis in dogs. The transmission of infection occurs by ingestion route. The algae pass through the intestinal mucosa and are disseminated throughout the body via blood or lymph. Gastrointestinal disorders producing vomition, tenesmus and intermittent diarrhoea (with blood and slime) are most commonly manifested. Diffuse hyperaemia, haemorrhages, ulcerations and multiple granulomas are observed in the intestinal mucosa (especially in the colon) which may result intestinal stricture and obstipation. The infection can spread to the central nervous system, cardiovascular system, urinary tract, liver, skeletal muscle, lymph nodes, thyroid gland, pancreas, peritoneum and diaphragm. The eye infection leading to blindness due to glaucoma and retinal ablation is another common clinical sign.

P. wickerhamii causes cutaneous protothecosis in cats and dogs. It is characterised by ulcerative lesions, scabs and pyogranulomatous dermatitis in the limbs, trunk and mucosal surfaces.

Human infection is primarily caused by *P. wickerhamii* (rarely by *P. zopfii*). Three clinical forms such as cutaneous, olecranon bursitis and systemic form are commonly detected. The cutaneous form occurs in immunocompromised individuals, and it is characterised by the formation of vesiculobullous and ulcerative lesion with purulent discharge, erythematous plaques, pustules, papules, nodules, pyodermic and herpetiform, and hypopigmented or atrophic lesions in exposed areas such as extremities and face. The olecranon bursitis takes place as a result of injuries or grazing of the elbow. The signs include mild induration of the bursa along with tenderness, erythema and production of serosanguineous fluid. The systemic form occurs in immunocompromised patients. The tissues and organs such as the skin, subcutaneous tissue, gut, peritoneum, blood and spleen are affected.

The clinical specimens for diagnosis of *Prototheca* infection in animals include faeces, urine, mastitic milk, skin scrapings and lesion exudates, whereas from human, skin scarificates (cutaneous protothecosis) or joint punctuate (olecranon bursitis) can be collected. Direct examination of the wet mount slides prepared from the clinical specimen or culture and stained with lactophenol cotton blue or calcofluor white can be done (Fig. 4.43). The tissues can be stained with Gridley fungus stain, Grocott's modification of Gomori methenamine silver, PAS with or without diastase and haematoxylin–eosin (less sensitive). *Prototheca* appears as large, spherical, oval or elliptical, non-budding cells both in wet mount and tissues. The size of *Prototheca* sporangia is usually smaller (10–30 μm) than the other fungi, and it contains 2–20 numbers of sporangiospores. In Gram-stained smears, spores stain as gram positive and the empty sporangia are gram negative. The PAS staining with diastase can differentiate *Prototheca* and green algae. The green algae contain cytoplasmic starch granules that are PAS negative following diastase treatment. Serological tests for the detection of anti-*Prototheca* antibodies can be performed in cattle to confirm the subclinical infection. The PCR based on 18S rDNA sequence is developed for differentiation of three *P. zopfii* genotypes. Taqman probe-based quantitative PCR is recently developed for the differentiation of *P. zopfii* genotype which is based on small subunit ribosomal DNA sequences.

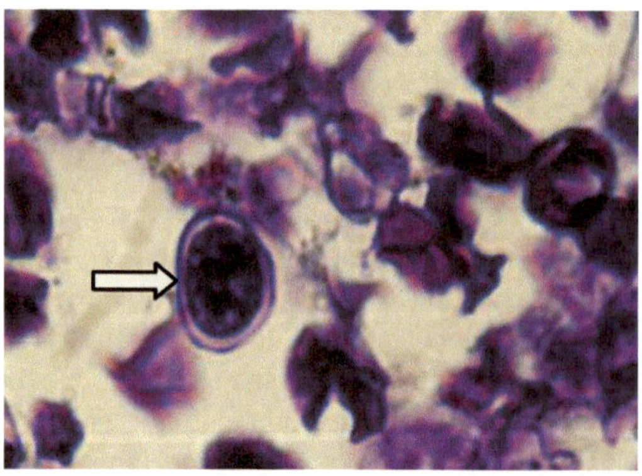

Fig. 4.43 *Prototheca* (1,000x) in PAS stained smear (Photograph courtesy: Prof. Cornelia Lass-Flörl, Medizinische Universität Innsbruck, Austria)

In human, amphotericin B is used in systemic protothecosis which prevents the ergosterol synthesis of *Prototheca*. The immidazoles have also attempted without any conclusive findings. Surgical excision is preferred in some cases. In immunocompromised individuals, the prognosis is grave.

In dogs with systemic protothecosis, amphotericin B and immidazoles are used, although the effect of such antifungals is not concluded. Surgical excision is performed in cutaneous form.

The effective treatment for protothecal bovine mastitis is yet not developed, although the in vitro antifungal sensitivity of *P. zopfii* was observed against amphotericin B, nystatin, polymyxin, gentamicin and neomycin. The control of mastitis can be achieved by detection and exclusion of infected animals from the herd and by adaptation of common hygienic measures such as teat dipping before and after milking, etc.

4.17.4 Lobomycosis

Lobomycosis is a chronic granulomatous fungal infection of the skin and subcutaneous tissues in human, bottlenose dolphin (*Tursiops truncatus*) and Guiana dolphin (*Sotalia guianensis*). The first human case was described by Jorge Lobo (1931) in Brazil. Lobo designated the infection as 'keloidal blastomycosis' due to the presence of organisms resembling *Paracoccidioides brasiliensis*. Later the disease was named Jorge Lobo's disease.

It is caused by *Lacazia loboi*, a yeast-like organism which is noncultivable in artificial culture medium. *Lacazia loboi* is taxonomically related (sister taxon) with *Paracoccidioides brasiliensis* and belongs to the order Onygenales. Both of the fungi share antigenic proteins (gp43, chitin synthase). The sequencing of *L. loboi* gp43 revealed its higher degree of nucleotide similarity (85 %) with *P. brasiliensis* gp43 and comparatively lower similarity (75 %) in amino acid level. Probably the high degree of nucleotide similarity occurs due to lack of known sexual reproduction and contact with its natural habitat for prolonged period (40–50 years) which helps to retain the original nucleotide motif in *L. loboi* genome. Further genomic study revealed about the loss of certain nucleotides (reductive evolution) which was earlier observed in other uncultivated, host-restricted pathogens such as *Mycobacterium leprae*, *Rickettsia prowazekii* and *Buchnera*. The reductive evolution occurs due to lack of known sexual cycle to acquire new DNA during recombination, and as a consequence decreased evolutionary fitness with every possibility of early extinction takes place.

Lobomycosis is endemic in South and Central Americas. The endemic zone should be located 200–250 m above sea level and should have dense vegetation, annual rainfall < 2,000 mm, mean temperature of 24 °C and mean relative humidity >75 %. However, the infection is also reported from Europe, Canada, the United States and South Africa. In this non-endemic zone, the persons who had travelled to Central or South America or had contact with an infected dolphin were detected with lobomycosis.

Transmission of *Lacazia loboi* in human occurs through traumatic inoculation such as scratches obtained while working in agriculture field and during insect and animal bites, etc. Lobomycosis in human is characterised by formation of cutaneous nodules, papules or plaques of various sizes having smooth, verrucous or ulcerated surfaces after a prolonged incubation period (more than 40 years). The lesions commonly appear in the extremities and ears which are sometimes disseminated into other body parts such as lymph nodes. The chronic lesions may be transformed into squamous cell carcinoma. In dolphins, lobomycosis is characterised by white to pink, verrucous lesions which may ulcerate and form large plaques.

Diagnosis depends on direct examination of the smears prepared from the clinical specimens (exudate) which show yeast-like rounded, thick-walled *Lacazia loboi* cells, occurring in chains of 2–10. The confirmation depends on histopathological examination showing atrophic epidermis and development of fibrous granuloma in the dermis. The granuloma is composed of histiocytes and giant cells containing the typical thick-walled *L. loboi* cells. The tissue sections are stained with periodic acid–Schiff, Gomori methenamine silver (GMS) and Gridley's silver stain to detect the yeast-like cells.

The localised lesions in the extremities are treated with cryosurgery or surgical excision in such a manner so that the margins of the lesions are free of infection to avoid recurrence. The disseminated lesions are treated with chemotherapy, such as clofazimine and itraconazole or a combination of both antifungals.

4.17.5 *Lagenidium*

Lagenidium giganteum was first described by Couch (1935) from copepods and mosquito larvae in North Carolina, USA. Grooters et al. (2000) first isolated and identified *Lagenidium* from a dog with severe multifocal cutaneous lesions and regional lymphadenomegaly.

The genus *Lagenidium* belongs to the class Oomycetes, and it contains more than 50 species. *Lagenidium* species are parasites of algae, fungi, rotifers, nematodes, crustaceans and mosquito larvae. *Lagenidium giganteum* is a facultative pathogen of mosquito larvae and is used as a biological control agent of mosquito (*Culex*, *Anopheles*, *Mansonia*). Canine pathogenic *Lagenidium* species are antigenically related with *L. giganteum*, although their life cycle and biology are unexplored.

Lagenidium possesses broad (7–25 μm in diameter), poorly septate or coenocytic, hyaline and branched hyphae. *L. giganteum* produces both asexual (zoospores) and sexual spores (oospores). The zoospores are biflagellate and motile which can infect the mosquito larvae. The zoospores do not have a cell wall and are fragile in nature (48 h survival period). The oospores are resistant to desiccation and mechanical abrasion and remain viable for several years in the environment.

L. giganteum was isolated from different parts of the world including North America, Europe, Africa, Asia and even Antarctica. Canine pathogenic *Lagenidium* is so far reported from young- to middle-aged dogs residing at south-eastern United States. They reside in the aquatic environment as a saprophyte. The dogs suffering from *Lagenidium* infection had previous exposures to ponds or lakes. *Lagenidium* are sterol auxotrophs and incorporate sterols from the environment rather than producing them.

Lagenidium can be isolated in peptone yeast glucose (PYG) agar with ampicillin (100 mcg/mL) and streptomycin (200 mcg/mL) and 2 % Sabouraud dextrose agar. The small pieces of tissues are directly placed on the surface of plates. The plates are incubated at 37 °C. The typical colonies are observed within 24–48 h. The colonies are submerged, glabrous and white to colourless. The zoospores can be artificially produced in Sabouraud dextrose agar (pH 7.0) and corn meal agar.

The infection produced by *Lagenidium* is known as lagenidiosis. The lagenidiosis in dogs is characterised by progressive cutaneous or subcutaneous multifocal lesions which involve the extremities, mammary gland, perineum and trunk. The lesions are nodular or ulcerated with draining tracts. The dissemination of infection occurs into the regional lymph nodes and abdominal great vessels. Sometimes the infection in dogs is associated with hyperglobulinaemia, hypoalbuminaemia and hypercalcaemia. Cats, ruminants and other domestic animals are not reported to be infected with *Lagenidium*. However, lagenidiosis is a common problem in shrimp culture, causing lethargy and mortality of larvae. In human, severe keratitis is reported with *Lagenidium* infection.

The diagnosis of lagenidiosis is based on histological observation, isolation of fungi and serological and molecular tests. In contrast to *Pythium*, the hyphae of *Lagenidium* can be observed in haematoxylin–eosin (H&E)-stained smear and the smears stained with Gomori methenamine silver (GMS). The *Lagenidium* hyphae are broader (7–25 μm) than *Pythium* hyphae (4–10 μm). The ELISA can detect the anti-*Lagenidium* antibodies in canine serum. However, these antibodies may produce cross-reaction with *Pythium* infection. So, *Lagenidium*-ELISA should always be performed along with *Pythium*-ELISA to get a confirmative diagnosis. The PCR is developed using the internal-transcribed spacer (ITS) of rRNA gene as a target. The PCR is successfully applicable for the confirmation of DNA extracted from the culture and clinical specimens if preserved properly. The sequencing of the PCR product and its alignment study can definitively diagnose the fungi.

In animals, surgical resection of infected tissues with wide margin is the treatment of choice for lagenidiosis. Amputation is suggested if the infection occurs in limbs. Assessment of internal organs is recommended before surgical resection to determine the stage of dissemination, if any. Treatment with antifungals is not so much successful. However, use of itraconazole and terbinafine combined with surgical resection was found effective in dogs. In human suffering from keratitis, repeat penetrating keratoplasty successfully eradicated the infection and prevented the recurrence.

4.17.6 *Basidiobolus* and *Conidiobolus*

Basidiobolus and *Conidiobolus* belong to the order Entomophthorales (entemon = insect) under the class Zygomycetes and phylum Zygomycota. Recently a new subphylum (*incertae sedis*), namely, Entomophthoromycotina, is proposed to include the order Entomophthorales. Major species under the genera are *Basidiobolus ranarum*, *Conidiobolus coronatus*, *C. incongruus* and *C. lamprauges*.

The hyphae are broad, thin walled, coenocytic with little branches. The hyphal diameter of *Basidiobolus* is 5–20 μm (mean 9 μm) and *Conidiobolus* is 5–13 μm (mean 8 μm). Both of the fungi produce asexual (conidia) and sexual spores (zygospores). The conidiophores are unbranched. Primary conidia are produced singly at the swollen tip of the conidiophores which are discharged forcibly. The primary conidia are spherical, 12–40 μm in diameter, and have a prominent basal papilla or hair-like projections (villose conidia). Secondary conidia are generated from the primary conidia which are smaller. Sometimes, oval to elongate spores with a terminal knob are produced which are known as 'capilliconidia'. The zygospores are thick-walled, smooth, undulant and contain a characteristic 'copulatory beak' (remnant of the copulation tube).

Most of them are saprophytes and are commonly found in soil, decaying plant materials. They are also present in insects and faeces of amphibians and reptiles. Both of the fungi are worldwide in distribution but are more prevalent in tropical countries or countries with warm and moist climate such as India, Australia, Africa, Central America, Brazil, Colombia and south-eastern United States.

Both of the fungi can be isolated in common media such as Sabouraud dextrose agar and potato dextrose agar. The plates are incubated at 25 °C for 2days (*Conidiobolus*) to 5 days (*Basidiobolus*). The typical colonies are flat, waxy in consistency and grey in colour. With the maturity, the colonies become heaped and radially folded. The aerial hyphae are generated and the different conidia are discharged. The discharged spores (primary, secondary and capilliconidia) cover the colony surface and lid of the Petri dish.

The transmission of the infection occurs by percutaneous entrance of the spores through the trauma, insect bite, etc. Mites and other insects can carry the fungal spores (capilliconidia of *Basidiobolus ranarum*) on their surface. Sometimes ingestion and inhalation of the spores can also transmit the infection. Dogs get infected while chewing of the wood which is old and decaying. The reptiles and amphibians excrete the spores in the environment, although the stability of the spores in the environment is low.

The virulence factors include thermotolerance, serine protease, lipase and collagenase. Thermotolerance (37 °C) helps in adaptation of the fungi in mammalian host. The serine protease is involved in discharge of fungal spores. Other proteases, lipase and collagenase cause tissue degradation and help in invasion of the fungi. Cellular degradation products provide nutrition to the fungi.

The major diseases produced by *Conidiobolus* and *Basidiobolus* are enlisted in Table 4.43.

The clinical specimens for the diagnosis of *Conidiobolus* and *Basidiobolus* infection in animals and human include polyp, scrapings,

Table 4.43 Major diseases of animal and human caused by *Conidiobolus* and *Basidiobolus*

Fungi	Host	Disease
Conidiobolus	Horses	*Equine conidiobolomycosis* (*rhinophycomycosis*): It is characterised by pyrogranulomatous lesion in nasal mucosa which causes epistaxis, nasal discharge and dyspnoea
	Sheep, deer, human	Nasopharyngeal infection with or without local dissemination into face, retropharyngeal region and retrobulbar space. Nasal polyp may develop which can block the passage. The human infection was previously described as 'subcutaneous phycomycosis'
	Dogs	*Nasopharyngeal conidiobolomycosis*: It is characterised by the formation of severe chronic ulcerative dermatitis in the nasal planum, ulcer in hard palate, regional lymphadenopathy, mutifocal ulcerative subcutaneous lesions and pneumonia
	Cats	Lesion in the gastrointestinal tract, lungs, hard palate
Basidiobolus	Dogs	Gastrointestinal tract lesion, focal preputial/vulvar lesion, ulcerative draining skin lesions (rare)
	Horses	Pruritic granulomatous lesions in the trunk, neck, thorax, ventral abdomen and head

exudates, nasal discharge and bronchoalveolar lavage. The direct examination of the specimens can be done with KOH mount which reveals the presence of broad, thin-walled, coenocytic hyphae with little branches in positive cases. The branches are emerged at right angles from the hyphae ('lateral peg'). The histological sections are characterised by granulomatous inflammation infiltrated with neutrophils, plasma cells, eosinophils and multinucleate giant cells. In haematoxylin–eosin-stained smear, the granuloma contains hyphae at the centre as clear or basophilic mass. It is surrounded by a wide (2.5–25 μm) eosinophilic sleeve ('eosinophilic cuff') which differentiates *Conidiobolus* and *Basidiobolus* infection (Entomophthorales infection) from others such as pythiosis and lagenidiosis. This sleeve is an antigen–antibody complex and the incident is called 'Splendore–Hoeppli phenomenon'. The serological test such as ELISA for the detection of antibodies is also developed.

The most effective treatment in animals is aggressive surgical resection of infected tissues which is followed by itraconazole for 2–3 months. If surgery is not preferred by the owners, the combination of itraconazole or amphotericin B lipid complex is recommended. The subcutaneous infection can be treated with itraconazole, amphotericin B lipid complex and potassium iodide. The equine conidiobolomycosis is also treated with direct injection of amphotericin B within the lesion along with systemic sodium and potassium iodide therapy.

4.17.7 Adiaspiromycosis (Adiaspirosis, Haplomycosis)

Adiaspiromycosis is a non-contagious, self-limiting pulmonary infection of mammals (rarely human) caused by dimorphic fungi of the genus *Emmonsia* (previously known as *Chrysosporium*). The fungus was first isolated by Emmons and Ashburn (1942) from wild rodents in Arizona. The fungus is currently worldwide in distribution and it is present as saprophyte in the soil. Due to production of *adiaconidia* by the fungi, the infection is known as adiaspiromycosis.

Emmonsia belongs to the family Onygenaceae. The major species under the genus are *Emmonsia crescens* (formerly *Chrysosporium parvum* var. *crescens*) and *E. parva*. The hyphae are hyaline, septate and branched. The conidiophores come up at right angle from the vegetative hyphae and produce small aleurioconidia (2–4 μm). These spores are transmitted into the host by inhalation. In the lung tissue (or in artificial culture at 40 °C), the spores are converted into a spherical structure (adiaconidia). The adiaconidium matures into a thick-walled spherule (200–400 μm in diameter). No further replication of spherule is observed. The spherule wall is refractile and bilayered. The outer layer is narrow and eosinophilic, whereas the inner layer is wide, hyaline and chiefly made of chitin. The spherules in the tissue appear as empty or may contain small eosinophilic globules. When the spherule concentration increases in the body, they may collapse the lung alveoli and produce respiratory distress and failure.

The adiaspiromycosis is detected in human, dogs, goat, horse and soil burrowing mammals such as armadillos, ground squirrels, mink, mice, skunks, etc. In human, three clinical forms are reported depending on the concentration of spores transmitted into the body. The forms are solitary granuloma, localised granulomatous form and disseminated form. In low concentration of spores, pulmonary nodules develop and the infection remains asymptomatic. In disseminated form, fever, cough and dyspnoea are observed due to compression and displacement of alveolar parenchyma by the expanding granulomas. Rarely the infection may spread into the other body parts such as the skin, peritoneum and bone. In animals, the infection is mostly asymptomatic and light grey to yellowish lesions are observed in the lungs.

The diagnosis depends on the histopathological detection of characteristic spherule in the lung tissue. In the tissues, the spherule is detected in the centre of a granuloma produced by giant cells, epithelioid cells and fibrous tissues. All the granulomas are in the same developmental stages, indicating that no replication of spherule occurs. The distinguishing thick wall and absence of sporangiospores help in differential diagnosis of *Emmonsia* spherules from others such as the spherules of *Coccidioides immitis* or *Rhinosporidium seeberi*. The fungi are difficult to isolate. Common fungal media such as Sabouraud dextrose agar without antibiotic are used (30 °C). The serological tests are not reliable for diagnosis. PCR and sequencing is developed for the confirmation of *Emmonsia crescens* from clinical specimens.

In immunocompetent patients, the infection is self-limiting and no treatment is required. In immunocompromised patients, the antifungals such as ketoconazole, fluconazole and amphotericin B are recommended.

Bibliography

Abad A, Fernndez-Molina JV, Bikandi J, Ramírez A, Margareto J, Sendino J, Hernando FL, Pontn J, Garaizar J, Rementeria A. What makes *Aspergillus fumigatus* a successful pathogen? Genes and molecules involved in invasive aspergillosis. Rev Iberoam Micol. 2010;27:155–82.

Abaineh D, Sintayehu A. Treatment trial of subclinical mastitis with the herb *Persicaria senegalense* (Polygonaceae). Trop Anim Health Prod. 2001;33:511–9.

Adenis AA, Aznar C, Couppi P. Histoplasmosis in HIV-infected patients: a review of new developments and remaining gaps. Curr Trop Med Rep. 2014;1:119–28.

Ahmad S, Khan Z. Invasive candidiasis: a review of nonculture-based laboratory diagnostic methods. Ind J Med Microbiol. 2012;30:264–9.

Ajello L. Natural history of the dermatophytes and related fungi. Mycopathol Mycol Appl. 1974;53:93–110.

Alby K, Bennett RJ. Sexual reproduction in the *Candida* clade: cryptic cycles, diverse mechanisms, and alternative functions. Cell Mol Life Sci. 2010;67:3275–85.

Alhuwalia KB, Maheshwari N, Deka RC. Rhinosporidiosis: a study that resolves etiologic controversies. Am J Rhinol. 1997;11:479–83.

Alibert J (1806) Description des malaides de la peau observe'es a'l'ho^pital Saint-Louis et exposition des meilleures me´thodes suivies pour leur traitement. Barrois, p 129

Almeida SR. Immunology of dermatophytosis. Mycopathol. 2008;166:277–83.

Andrews MD, Burns M. Common tinea infections in children. Am Fam Physician. 2008;77:1415–20.

Arseculeratne SN. Recent advances in rhinosporidiosis and *Rhinosporidium seeberi*. Indian J Med Microbiol. 2002;20:119–31.

Ashford B. Certain conditions of the gastrointestinal tract in Puerto Rico and their relation to tropical sprue. Am J Trop Med Hyg. 1928;8:507–38.

Ashman RB, Papadimitriou JM, Olt AK, Warmttigton JR. Antigens and immune responses in *Candida albicans* infection. Immunol Cell Biol. 1990;68:1–13.

Ashworth JH. On *Rhinosporidium seeberi* (Wernicke, 1903) with special reference to its sporulation and affinities. Trans Roy Soc Edin. 1923;53:301–42.

Asthana RP. Aspergillosis in fowls. Proc Indian Acad Sci Sec-B. 1944;20:43–7.

Austwick PKC, Copland JW. Swamp cancer. Nature. 1974;250:84.

Ayyar VK. Rhinosporidiosis in cattle. Trans Far-East Assoc Trop Med. 1927;3:658–64.

Baker RD. Mucormycosis; a new disease? J Am Med Assoc. 1957;163:805–8.

Bariola JR, Vyas KS. Pulmonary blastomycosis. Semin Respir Crit Care Med. 2011;32:745–53.

Barros MBL, Schubach AO, do Valle AC, et al. Cat-transmitted sporotrichosis epidemic in Rio de Janeiro, Brazil: description of a series of cases. Clin Infect Dis. 2004;38:529–35.

Barros MBL, Paes RDA, Schubach AO. *Sporothrix schenckii* and sporotrichosis. Clin Microbiol Rev. 2011;24:633–54.

Barton RC. Laboratory diagnosis of invasive aspergillosis: from diagnosis to prediction of outcome. Scientifica (Cairo). 2013;2013:459405.

Barua HK, Ahmed I. Experimental studies on mycotic infection in cattle. Indian Agric. 1963;7:139–46.

Bauer H, Ajello L, Adams E, Hernandez DU. Cerebral mucormycosis: pathogenesis of the disease. Am J Med. 1955;18:822–31.

Beernaert LA, Pasmans F, Van Waeyenberghe L, Haesebrouck F, Martel A. *Aspergillus* infections in birds: a review. Avian Pathol. 2010;39:325–31.

Beguin H, Goens K, Hendrickx M, Planard C, Stubbe D, Detandt M. Is *Trichophyton simii* endemic to the Indian subcontinent? Med Mycol. 2013;51:444–8.

Ben-Ami R, Lewis RE, Kontoyiannis DP. Enemy of the (immunosuppressed) state: an update on the pathogenesis of *Aspergillus fumigatus* infection. Br J Haematol. 2010;150:406–17.

Bergmans AM, van der Ent M, Klaassen A, Böhm N, Andriesse GI, Wintermans RG. Evaluation of a single-tube real-time PCR for detection and identification of 11 dermatophyte species in clinical material. Clin Microbiol Infect. 2010;16:704–10.

Bialek R, Kern J, Herrmann T, Tijerina R, Ceceñas L, Reischl U, Gonzlez GM. PCR assays for identification of *Coccidioides posadasii* based on the nucleotide sequence of the antigen 2/proline-rich antigen. J Clin Microbiol. 2004;42:778–83.

Bianchi M, Robles M, Vitale R, Helou S, Arechavala A, Negroni R. The usefulness of blood cultures in the diagnosis of AIDS-related systemic mycosis: evaluation of a manual lysis-centrifugation method. Medical Mycol. 2000;38:77–80.

Binnicker MJ, Buckwalter SP, Eisberner JJ, Stewart RA, McCullough AE, Wohlfiel SL, Wengenack NL. Detection of *Coccidioides* species in clinical specimens by real-time PCR. J Clin Microbiol. 2007;45:173–8.

Bonifaz A, Vzquez-Gonzlez D. Sporotrichosis: an update. Gital Dermatol Venereol. 2010;145:659–73.

Bonifaz A, Vzquez-Gonzlez D. Diagnosis and treatment of lymphocutaneous sporotrichosis: what are the options? Curr Fungal Infect Rep. 2013;7:252–9.

Boro BR, Chakrabarty AK, Sarma G, Sarmah AK. Ringworm in animals due to *Epidermophyton floccosum*. Vet Rec. 1980;107:491–2.

Bourgeois C, Kuchler K. Fungal pathogens: a sweet and sour treat for *toll* like receptors. Frontiers Cell Infect Microbiol. 2012;2:1–11.

Bovers M, Hagen F, Boekhout T. Diversity of the *Cryptococcus neoformans-Cryptococcus gattii* species complex. Rev Iberoam Micol. 2008;25:S4–12.

Bradford K, Meinkoth J, McKeirnen K, Love B. Candida peritonitis in dogs: report of 5 cases. Vet Clin Pathol. 2013;42:227–33.

Bradsher RW. Histoplasmosis and blastomycosis. Clin Infect Dis. 1996;22:S102–10.

Bradsher RW, Chapman SW, Pappas PG. Blastomycosis. Infect Dis Clin N Am. 2003;17:21–40.

Brasch J, Horter F, Fritsch D, Beck-Jendroschek V, Tröger A, Francke W. Acyclic sesquiterpenes released by *Candida albicans* inhibit growth of dermatophytes. Med Mycol. 2014;52:46–55.

Brillowska-Dabrowska A, Swierkowska A, Lindhardt Saunte DM, Arendrup MC. Diagnostic PCR tests for *Microsporum audouinii*, *M. canis* and *Trichophyton* infections. Med Mycol. 2010;48:486–90.

British Medical Research Council (1977) cited by Fahal AH. Mycetoma: a thorn in the flesh. Trans R Soc Trop Med Hyg. 2004;98:3–11.

Brumpt (1905) cited by Mackinnon JE. A contribution to the study of the causal organisms of maduromycosis. Trans R Soc Trop Med Hyg. 1954;48:470–72.

Brumpt E. Les mycetomes. Arch Parasitol. 1906;10:489–564.

Brunke S, Hube B. Two unlike cousins: *Candida albicans* and *C. glabrata* infection strategies. Cell Microbiol. 2013;15:701–8.

Busse O. Uber parasitare Zelleinschlusse und ihre Zuchtung. Centralbl Bakt Parasit. 1894;16.

Butler G. Fungal sex and pathogenesis. Clin Microbiol Rev. 2010;23:140–59.

Cafarchia C, Iatta R, Latrofa MS, Graser Y, Otranto D. Molecular epidemiology, phylogeny and evolution of dermatophytes. Infect Genet Evol. 2013;20:336–51.

Cantrell SA, Dianese JC, Fell J, Gunde-Cimerman N, Zalar P. Unusual fungal niches. Mycologia. 2011;103:1161–74.

Capoor MR, Sen B, Varshney P, Verghese M, Shivaprakash MR, Chakrabarti A. Coccidioidomycosis masquerading as skeletal tuberculosis: an imported case and review of coccidioidomycosis in India. Trop Doct. 2014;44:25–8.

Capponi M, Sureau P, Segretain G. Pnicillose de *Rhizomys sinensis*. Bull Soc Pathol Exot. 1956;49:418–21.

Carmichael JW. *Chrysosporium* and some other aleuriosporic Hyphomycetes. Can J Bot. 1962;40:1137–73.

Carr J, Shearer G. Genome Size, complexity, and ploidy of the pathogenic fungus *Histoplasma capsulatum*. J Bacteriol. 1998;180:6697–703.

Carter H. On a new striking form of fungus disease principally affecting the foot and prevailing endemically in many parts of India. Trans Med Phys Soc Bombay. 1860;60:140–2.

Carter HV. On mycetoma or the fungs disease of India. London: J. & A Churchill; 1874.

Chaffin WL. *Candida albicans* cell wall proteins. Microbiol Mol Biol Rev. 2008;72:495–544.

Chagas C. Nova tripanozomiazaea humana. Mem Inst Oswaldo Cruz. 1909;1:159–218.

Chakrabarti A. Epidemiology of central nervous system mycosis. Neurol India. 2007;55:191–7.

Chakrabarti A, Singh R. The emerging epidemiology of mould infections in developing countries. Curr Opin Infect Dis. 2011;24:521–6.

Chakrabarti A, Slavin MA. Endemic fungal infections in the Asia-Pacific region. Med Mycol. 2011;49:337–44.

Chakrabarti A, Jatana M, Kumar P, Chatha L, Kaushal A, Padhye A. Isolation of *Cryptococcus neoformans* var. *gattii* from *Eucalyptus camaldulensis* in India. J Clin Microbiol. 1997;35:3340–2.

Chakrabarty AN, Ghosh S, Blank F. Isolation of *Trichophyton rubrum* (Castollani) Sab., 1911 from animals. Can J Comp Med. 1954;18:436–8.

Chandler FW, Kaplan W, Ajello L. A color atlas and textbook of the histopathology of mycotic diseases. Chicago: Year Book Medical Publishers, Inc; 1980. p. 104–5.

Chaturvedi V, Chaturvedi S. *Cryptococcus gattii*: a resurgent fungal pathogen. Trends Microbiol. 2011;19:564–71.

Chauhan HVS, Sadana JR. Gastric phycomycosis in a piglet. Indian Vet J. 1973;50:1150–6.

Chauhan HVS, Sharma GL, Kalra DS, Malhotra FC, Kapur MP. A fatal cutaneous granuloma due to *Entomophthora coronata* in a mare. Vet Rec. 1972;92:425.

Chhabra D, Moghe MN, Tanwani SK, Sharada R. Mycotic mastitis in buffaloes. Ind J Comp Microbiol Immunol Infect Dis. 1998;19:108–9.

Chick EW, Evans J, Baker RR. The inhibitory effect of amphotericin B on localized *Rhizopus oryzae* infection (mucormycosis) utilizing the pneumoderma pouch of the rat. Antibiot Chemother. 1958;8:506–10.

Chitravel V, Sundararaj T, Subramanian S, Kumaresan M, Kunjithapadam S. Detection of circulating antigen in patients with rhinosporidiosis. Sabouraudia. 1982;20:185–91.

Chodat R. Monographies d'algues en culture pure. Mat Crypt Suisse. 1913;4:234–41.

Chooi YH, Wang P, Fang J, Li Y, Wu K, Wang P, Tang Y. Discovery and characterization of a group of fungal polycyclic Polyketide Prenyltransferases. J Am Chem Soc. 2012;134:9428–37.

Clark BM. Epidemiology of phycomycosis. In: Wolstenhome GEW, Porter R, editors. Systemic mycoses. Boston: Little, Brown & Co; 1968. p. 179–92.

Coelho LM, Cursino-Santos JR, Persinoti GF, Rossi A, Martinez-Rossi NM. The *Microsporum canis* genome is organized into five chromosomes based on evidence from electrophoretic karyotyping and chromosome end mapping. Med Mycol. 2013;51:208–13.

Connolly P, Hage CA, Bariola JR, Bensadoun E, Rodgers M, Bradsher RW, Wheat LJ. *Blastomyces dermatitidis* antigen detection by quantitative enzyme immunoassay. Clin Vac Immunol. 2012;19:53–6.

Cortez KJ, Roilides E, Quiroz-Telles F, Meletiadis J, Antachopoulos C, Knudsen T, Buchanan W, Milanovich J, Sutton DA, Fothergill A, Rinaldi MG, Shea YR, Zaoutis T, Kottilil S, Walsh TJ. Infections caused by *Scedosporium* spp. Clin Microbiol Rev. 2008;21:157–97.

Costa EQ, Diniz LS, Netto CF, Arruda C, Dagli ML. Epidemiological study of sporotrichosis and histoplasmosis in captive Latin American wild mammals, São Paulo, Brazil. Mycopathologia. 1994;125:19–22.

Couch JN. A new saprofytic species of *Coelomomyces*, with notes on other forms. Mycologia. 1935;27:376–87.

Cox RA, Magee DM. Coccidioidomycosis: host response and vaccine development. Clin Microbiol Rev. 2004;17:804–39.

Curtis F. Contribution à l'tude de la saccharomycose humaine. Ann Inst Pasteur. 1896;10:449–68.

Dagenais TRT, Keller NP. Pathogenesis of *Aspergillus fumigatus* in invasive Aspergillosis. Clin Microbiol Rev. 2009;22:447.

Damodaran S, Ramachandran PV, Thainkachalam M, Mahalingam P. Mycotic gastritis in cattle. Indian Vet J. 1976;43:848–51.

Darling ST. A protozoon general infection producing pseudotubercles in the lungs and focal necroses in the liver, spleen and lymphnodes. J Am Med Asso. 1906;46:1283–5.

Das PK, Joseph E. Identification and antibigram of microbes associated with buffalo mastitis in Jabalpur, Madhya Pradesh, India. Buffalo Bull. 2005;24:3–9.

Datta SCA. Chronic bovine haematuria. Rep. 1st Imp. Vet. Conf. Lond. Weybridge: Imperial Bureau of Animal Health; 1938. p. 78–9.

Davies RR, Spencer H, Wakelin PO. A case of human protothecosis. Trans R Soc Trop Med Hyg. 1964;58:448–51.

De Cock AW, Mendoza L, Padhye AA, Ajello L, Kaufman L. *Pythium insidiosum* sp. nov., the etiologic agent of pythiosis. J Clin Microbiol. 1987;25:344–9.

De Haan J, Hoogkamer LJ. Hyphomycosis destruens. Veeartsenijkundige bladen voor Nederlandsh-Indie. 1901;13:350–74.

de Klerk N, de Vogel C, Fahal A, van Belkum A, van de Sande WW. Fructose-bisphosphate aldolase and pyruvate kinase, two novel immunogens in Madurella mycetomatis. Med Mycol. 2012;50:143–51.

De Monbreun WA. The cultivation and cultural characteristics of Darling's *Histoplasma capsulatum*. Am J Trop Med. 1934;14:93–125.

De Monbreun WA. The dog as a natural host for *Histoplasma capsulatum*: report of a case of histoplasmosis in this animal. Am J Trop Med. 1939;19:565–87.

de Souza W, Benchimol M. Basic biology of *Pneumocystis carinii* – a mini review. Mem Inst Oswaldo Cruz. 2005;100:903–8.

Deepe GS. Immune response to early and late *Histoplasma capsulatum* infections. Curr Opin Microbiol. 2000;3:359–62.

Dey NC, Kakoti LM. *Microsporum gypseum* in India. J Indian Med Assoc. 1955;25:160–4.

DiCaudo DJ. Coccidioidomycosis: a review and update. J Am Acad Dermatol. 2006;55:929–42.

Dickson E. "Valley fever" of the San Joaquin Valley and fungus Coccidioides. Calif West Med. 1937;47:151–7.

Difonzo EM, Cappugi P, Moretti S, Panconesi E. Kerion-like *tinea barbae* caused by *Epidermophyton floccosum*. Mykosen. 1985;28:365–8.

DiSalvo AF. Infection caused by *Penicillium marneffei*: a description of first natural infection in man. Am J Clin Pathol. 1973;60:259–63.

Dyer PS, O'Gorman CM. A fungal sexual revolution: *Aspergillus* and *Penicillium* show the way. Curr Opin Microbiol. 2011;14:1–6.

Ehrenberg CF. De Mycetogenesi. Nov Act Acad Leop. 1820;11:198.

Ellepola ANB, Morrison CJ. Laboratory diagnosis of Invasive Candidiasis. J Microbiol. 2005;43:65–84.

Ellis DH, Pfeiffer TJ. Natural habitat of *Cryptococcus neoformans* var. *gattii*. J Clin Microbiol. 1990;28:1642–4.

Emmons CW. Dermatophytes. Natural grouping based on the form of the spores and accessory organs. Arch Dermatol Syphilol. 1934;30:337–62.

Emmons CW. Isolation of *Cryptococcus neoformans* from soil. J Bacteriol. 1951;62:685–90.

Emmons CW. *Cryptococcus neoformans* strains from a severe outbreak of bovine mastitis. Mycopathol Mycol Appl. 1952;6:23–4.

Emmons CW, Ashburn LL. The isolation of *Haplosporangium parvum* n. sp. and *Coccidioides immitis* from wild rodents. Public Hlth Rep. 1942;57:1715–27.

Ernst HC. A case of mucor infection. J Med Res. 1918;39:143–6.

Fon TE, Bah MM, Yongbang BW, Tata GN, Ambe AE. Antimycotic activity of *Commelina cyanea*. Vetscan. 2013;7:31–5.

Fredricks DN, Jolley JA, Lepp PW, Kosek JC, Relman DA. *Rhinosporidium seeberi:* a human pathogen from a novel group of aquatic Protistan parasites. Emerg Infect Dis. 2000;6:273–82.

Gaastra W, Lipman LJA, De Cock AWAM, Exel TK, Pegge RBG, Scheurwater J, Vilela R, Mendoza L. *Pythium insidiosum*: an overview. Vet Microbiol. 2010;146:1–16.

Ganguly P. Blastomycosis in Duars. Ind Med Gaz. 1925;60:189.

Gatti F, Eeckels R. An atypical strain of *Cryptococcus neoformans* (San Felice) Vuillemin 1894. I. Description of the disease and of the strain. Ann Soc Belges Med Trop Parasitol Mycol. 1970;50:689–93.

Gauglitz GG, Callenberg H, Weind G, Korting HC. Host defence against *Candida albicans* and the role of pattern recognition receptors. Acta Derm Venereol. 2012;92:291–8.

Gaunt MC, Taylor SM, Kerr ME. Central nervous system blastomycosis in a dog. Can Vet J. 2009;50:959–62.

Gavanji S, Asgari MJ, Vaezi R, Larki B. Antifungal effect of the extract of propolis on the growth of three species of *Epidermophyton flucosum, Trichophyton violaseum* and *Trichophyton tonsorans* in laboratory environment. African J Pharmacy Pharmacol. 2011;5:2642–6.

Gentles JC. Experimental ringworm in guinea pigs: oral treatment with griseofulvin. Nature. 1958;182:476–7.

Gilchrist TC. Protozoan dermatitis. J Cutan Gen Dis. 1894;12:496–9.

Gilchrist TC, Stokes WR. A case of pseudolupus caused by *Blastomyces*. J Exp Med. 1898;3:53–78.

Giri S, Kindo AJ. A review of *Candida* species causing blood stream infection. Ind J Med Microbiol. 2012;30:270–8.

Godfrey J. Diseases of the foot not hitherto described. Lancet. 1846;i:593–4.

Gonzlez A. Innate immune response to the dimorphic fungal pathogen *Coccidioides*: molecular and cellular mechanisms. J Clin Cell Immunol. 2013;S13:001.

Gow NAR, van de Veerdonk FL, Brown AJP, Netea MG. *Candida albicans* morphogenesis and host defence: discriminating invasion from colonization. Nat Rev Microbiol. 2012;10:112–22.

Gregory JE, Golden A, Haymaker W. Mucormycosis of the central nervous system. A report of three cases. Bull Johns Hopkins Hosp. 1943;73:405–19.

Griffin DN. The re-discovery of *Gymnoascus gypseum*, the perfect state of *Microsporum gypseum:* and a note on *Trichophyton terrestre*. Trans Br Mycol Soc. 1960;43:637–41.

Grooters AM. Pythiosis, lagenidiosis, and zygomycosis in small animals. Vet Clin Small Anim. 2003;33:695–720.

Grooters AM, Hodgin EC, Bauer RW, et al. *Lagenidium* sp infection in six dogs with subcutaneous and systemic disease: initial description of an emerging oomycosis. 10th annual focus on fungal infections; Atlanta, GA; March 29–31, 2000.

Gruby D. Recherches sur la nature, la siege et le developpement du porrigo dicalvans ou phytoalopecie. CR Acad Sci (París). 1843;17:301–3.

Gryganskyi AP, Lee SC, Litvintseva AP, Smith ME, Bonito G, et al. Structure, function, and phylogeny of the mating locus in the *Rhizopus oryzae* complex. PLoS One. 2010;5:e15273.

Guedes HLM, Guimarães AJ, Muniz MM, Pizzini CV, Hamilton AJ, Peralta JM, Deepe GS, Zancop-Oliveira RM. PCR Assay for identification of *Histoplasma capsulatum* based on the nucleotide sequence of the M Antigen. J Clin Microbiol. 2003;41:535–9.

Gugnani HC. Histoplasmosis in Africa: a review. Ind J Chest Dis Allied Sci. 2000;42:271–7.

Gumaa SA, Mahgoub ES. Counter immunoelectrophoresis in the diagnosis of mycetoma and its sensitivity as compared to immunodiffusion. Sabouraudia. 1975;13:309–15.

Gupta PK, Singh RP, Singh IP. Dermatophytes from man, dogs and pigs with special reference to *Trichophyton simii* and *Microsporum nanum*. Indian J Anim Hlth. 1968;11:247–53.

Gupta PK, Singh RP, Singh IP. A study of dermatomycosis (ringworm) in domestic animals and fowls. Indian J Anim Hlth. 1970;9:85–9.

Gyawali R, Lin X. Mechanisms of uniparental mitochondrial DNA inheritance in *Cryptococcus neoformans*. Mycobiology. 2011;39:235–42.

Harz OC. Einige neue Hyphomyceten Berlin's and Wien's nebst Beiträgen zur Systematik derselben. Bull Soc Imp Nat Moscou. 1870;44:87–147.

Haselow DT, Safi H, Holcomb D, Smith N, Wagner KD, Bolden BB, Harik NS. Arkansas Department of Health. Histoplasmosis associated with a bamboo bonfire – Arkansas, October 2011. MMWR Morb Mortal Wkly Rep. 2014;63:165–8.

Hayashi T, Sugita T, Hata E, Katsuda K, Zhang E, Kiku Y, Sugawara K, Ozawa T, Matsubara T, Ando T, Obayashi T, Ito T, Yabusaki T, Kudo K, Yamamoto H, Koiwa M, Oshida T, Tagawa Y, Kawai K. Molecular-based identification of yeasts isolated from bovine clinical mastitis in Japan. J Vet Med Sci. 2013;75:387–90.

Hektoen L, Perkins CF. Refractory subcutaneous abscesses caused by *Sporothrix schenckii*. A new pathogenic fungus. J Exp Med. 1900;5:77–89.

Hogan LH, Klein BS, Levitz SM. Virulence factors of medically important fungi. Clin Microbiol Rev. 1996;9:469–88.

Holzworth J. Cryptococcosis in a cat. Cornell Vet. 1952;42:12–5.

Houbraken J, de Vries RP, Samson RA. Modern taxonomy of biotechnologically important *Aspergillus* and

Penicillium species. Adv Appl Microbiol. 2014;86:199–249.

Howard DH. Dimorphism of *Sporotrichum schenckii*. J Bacteriol. 1961;81:464–9.

Hung C, Xue J, Cole GT. Virulence mechanisms of *Coccidioides*. Ann NY Acad Sci. 2007;1111:225–35.

Ibrahim AS, Spellberg B, Walsh TJ, Kontoyiannis DP. Pathogenesis of mucormycosis. Clin Infect Dis. 2012;54:S16–22.

Iyer PKR. Pulmonary blastomycosis in a dog in India. Ind J Vet Pathol. 1983;7:60–2.

Jain N, Guerrero A, Fries BC. Phenotypic switching and its implications for the pathogenesis of *Cryptococcus neoformans*. FEMS Yeast Res. 2006;6:480–8.

Jiang N, Yang Y, Janbon G, Pan J, Zhu X. Identification and functional demonstration of miRNAs in the fungus *Cryptococcus neoformans*. PLoS One. 2012;7:e52734.

Johnson SM, Simmons KA, Pappagianis D. Amplification of coccidioidal DNA in clinical specimens by PCR. J Clin Microbiol. 2004;42:1982–5.

Kaithwas G, Mukerjee A, Kumar P, Majumdar DK. *Linum usitatissimum* (linseed/flaxseed) fixed oil: antimicrobial activity and efficacy in bovine mastitis. Inflammopharmacol. 2011;19:45–52.

Kamiya A, Kikuchi A, Tomita Y, Kanbe T. PCR and PCR-RFLP techniques targeting the DNA topoisomerase II gene for rapid clinical diagnosis of the etiologic agent of dermatophytosis. J Dermatol Sci. 2004;34:35–48.

Kandhari KC, Sethi KK. Dermatophytosis in Delhi area. J Indian Med Assoc. 1964;42:324–6.

Kano R, Fujino Y, Takamoto N, Tsujimoto H, Hasegawa A. PCR detection of the *Cryptococcus neoformans* CAPS9 gene from a biopsy specimen from a case of feline cryptococcosis. J Vet Diagn Invest. 2001;13:439–42.

Kauffman CA. Histoplasmosis: a clinical and laboratory update. Clin Microbiol Rev. 2007;20:115–32.

Khan ZU, Randhawa HS, Lulla M. Isolation of *Blastomyces dermatitidis* from the lungs of a bat *Rhinopoma hardwickei hardwickei* gray in Delhi. Sabouradia. 1982;20:137–44.

Kini S, Pal D, Kowshik T, Randhawa HS, Deshpande S, Khan ZU. Sporotrichosis in India: first authentic case report from the north-western region and a critical literature review. J Med Vet Mycol. 1986;24:289–95.

Kirkland TN, Fierer J. Coccidioidomycosis: a reemerging infectious disease. Emerg Infect Dis. 1996;3:193–9.

Kirsch EJ, Greene RT, Prahl A, Rubin SI, Sykes JE, Durkin MM, Wheat LJ. Evaluation of *Coccidioides* antigen detection in dogs with Coccidioidomycosis. Clin Vaccine Immunol. 2012;19:343–5.

Klein BS, Jones JM. Isolation, purification, and radiolabeling of a novel 120-kD surface protein on *Blastomyces dermatitidis* yeasts to detect antibody in infected patients. J Clin Invest. 1990;85:152–61.

Klein BS, Jones JM. Purification and characterization of the major antigen WI-1 from *Blastomyces dermatitidis* yeasts and immunological comparison with A antigen. Infect Immun. 1994;62:3890–900.

Klein BS, Tebbets B. Dimorphism and virulence in fungi. Curr Opin Microbiol. 2007;10:314–9.

Kligman AM. The pathogenesis of tinea capitis due to *Microsporum audouini* and *Microsporum canis*. I. Gross observations following the inoculation of humans. J Invest Dermatol. 1952;18:231–46.

Kozubowski L, Heitman J. Profiling a killer, the development of *Cryptococcus neoformans*. FEMS Microbiol Rev. 2012;36:78–94.

Krishnan K, Askew DS. A regulatory hub for virulence traits in the mold pathogen *Aspergillus fumigatus*. Virulence. 2014;5:1–7.

Kroetza DN, Deepe GS. The role of cytokines and chemokines in *Histoplasma capsulatum* infection. Cytokine. 2012;58:112–7.

Krüger W. Kurz charakteristik einiger niederer organismen in saftflusse der laubbäume. I. Über einen neuen pilz-typus, repräsentiert durch die gattung Prototheca (Pr. Muriformis et P. zopfii). Hedwigia. 1894;33:241–51.

Kumar P, Yang M, Haynes BC, Skowyra ML, Doering TL. Emerging themes in cryptococcal capsule synthesis. Curr Opin Struct Biol. 2011;21:597–602.

Kunert J. Effect of reducing agents on proteolytic and keratinolytic activity of enzymes of Microsporumgypseum. Mycoses. 1992;35:343–8.

Lam WK. Lung cancer in Asian women-the environment and genes. Respirol. 2005;10:408–17.

Laniado-Laborín R. Coccidioidomycosis and other endemic mycoses in Mexico. Rev Iberoam Micol. 2007;24:249–58.

Laveran MA. Au sujet d'un cas de mycetoma à grains noirs. Bull Acad Med Paris. 1902;47:773–6.

Leach MD, Brown AJP. Posttranslational modifications of proteins in the pathobiology of medically relevant fungi. Eukaryot Cell. 2012;11:98–108.

Leal SM, Pearlman E. The role of cytokines and pathogen recognition molecules in Fungal Keratitis – insights from human disease and animal models. Cytokine. 2012;58:107–11.

Lerch M. Einen durch Algen (*Prototheca*) hervorgerufene Mastitis der Kuh. Berl Münch Tierärztl Wschr. 1952;65:64–9.

Li CH, Cervantes M, Springer DJ, Boekhout T, Ruiz-Vazquez RM, Torres-Martinez SR, Heitman J, Lee SC. Sporangiospore size dimorphism is linked to virulence of *Mucor circinelloides*. PLoS Pathog. 2011;7:e1002086.

Li W, Sullivan TD, Walton E, Averette AV, Sakthikumar S, Cuomo CA, Klein BS, Heitmana J. Identification of the Mating-Type (*MAT*) Locus that controls sexual reproduction of *Blastomyces dermatitidis*. Eukaryot Cell. 2013;12:109–17.

Link HF. Observationes in Ordines plantarum naturales, Dissertatio 1ma (Berlin Ges. NatKde31-42), Berlin; 1809.

Liu TB, Perlin DS, Xue C. Molecular mechanisms of cryptococcal meningitis. Virulence. 2012;3:173–81.

Lobo J O. Um caso de blastomicose produzido por uma espécie nova, encontrada em Recife. Rev Med Pernamb. 1931;1:763–5.

Loftus BJ, Fung E, Roncaglia P, Rowley D, Amedeo P, Bruno D, Vamathevan J, Miranda M, Anderson IJ, Fraser JA, Allen JE, Bosdet IE, Brent MR, Chiu R, Doering TL, Donlin MJ, D'Souza CA, Fox DS, Grinberg V, et al. The genome of the basidiomycetous yeast and human pathogen *Cryptococcus neoformans*. Science. 2005;307:1321–4.

Lohse MB, Johnson AD. White-opaque switching in *Candida albicans*. Curr Opin Microbiol. 2009;12:650–4.

Lopes-Bezerra LM. *Sporothrix schenckii* cell wall peptidorhamnomannans. Frontiers Microbiol. 2011;2:1–4.

Lortholary O, Denning DW, Dupont B. Endemic mycoses: a treatment update. J Antimicrob Chemother. 1999;43:321–31.

Lund A, Bratberg AM, Nass B, Gudding R. Control of bovine ringworm by vaccination in Norway. Vet Immunol Immunopathol. 2014;158:37–45.

Lutz A, Splendore A. Sobre uma mycose observada em homens e ratos. Rev Med São Paulo. 1907;21:433–50.

Macêdo DPC, Neves RP, Magalhães OMC, de Souza-Motta CM, de Queiroz LA. Pathogenic aspects of *Epidermophyton floccosum* langeron et milochevitch as a possible aethiological agent of *tinea capitis*. Brazilian J Microbiol. 2005;36:36–7.

Mahmoudabadi AZ, Zarrin M. Mycetomas in Iran: a review article. Mycopathol. 2008;165:135–41.

Maiti PK, Ray A, Bandyopadhyay S. Epidemiological aspects of mycetoma from a retrospective study of 264 cases in West Bengal. Trop Med Int Health. 2002;7:788–92.

Malbram (1892) cited by Seeber GR. Rhinosporidium kinealyi et Rhinosporidium seeberi une question de priorité Buenos Aires, Argentina: La Ciencia Médica. 1912. p. 10.

Malik R, Krockenberger MB, Cross G, Doneleyy R, Madillz DN, Black D, Mcwhirter P, Rozenwax A, Rose K, Alleyyy M, Forshawzz D, Russell-Brown I, Johnstoneyy AC, Martin P, O'brien CR, Love DN. Avian cryptococcosis. Med Mycol. 2003;41:115–24.

Malmsten PH. *Trichophyton tonsurans*, translated from 1845 Swedish paper by F. C. H. Creplin. Arch f Anat Physiol u wissensch Med. 1948;15:1–19.

Mansour MK, Tam JM, Vyas JM. The cell biology of the innate immune response to *Aspergillus fumigatus*. Ann NY Acad Sci. 2012;1273:78–84.

Mao L, Zhang L, Li H, Chen W, Wang H, Wu S, Guo C, Lu A, Yang G, An L, Abliz P, Meng G. Pathogenic fungus *Microsporum canis* activates the NLRP3 inflammasome. Infect Immun. 2014;82:882–92.

Martagon-Villamil J, Shrestha N, Sholtis M, Isada CM, Hall GS, Byrne T, Lodge BA, Reller LB, Procop GW. Identification of *Histoplasma capsulatum* from culture extracts by real-time PCR. J Clin Microbiol. 2003;41:1295–8.

Mayer FL, Wilson D, Hube B. *Candida albicans* pathogenicity mechanisms. Virulence. 2013;4:119–28.

McKinnell JA, Pappas PG. Blastomycosis: new insights into diagnosis, prevention, and treatment. Clin Chest Med. 2009;30:227–39.

Mendoza L, Vilela R, Rosa PS, Belone AFF. *Lacazia loboi* and *Rhinosporidium seeberi:* a genomic perspective. Rev Iberoam Micol. 2005;22:213–6.

Minchin EA, Fantham HB. *Rhinosporidium kinealyi*, ng., n. sp., a new sporozoon from the mucous membrane of the septum nasi of man. QJ Microsc Sci. 1905;49:521–32.

Monga DP, Kalra DS. Prevalence of mycotic mastitis among animals in Hariana. Indian J Anim Sci. 1971;41:813–6.

Monod M. Secreted proteases from dermatophytes. Mycopathol. 2008;166:285–94.

Montagu (1813) cited by Ainsworth GC. Introduction to the history of medical and veterinary mycology. Cambridge: Cambridge University Press; 1986.

Moran GP, Coleman DC, Sullivan DJ. Comparative genomics and the evolution of pathogenicity in human pathogenic fungi. Eukaryot Cell. 2011;10:34–42.

Moriello KA, DeBoer DJ, Greek J, Kuhl K, Fintelman M. The prevalence of immediate and delayed type hypersensitivity reactions to *Microsporum canis* antigens in cats. J Feline Med Surg. 2003;5:161–6.

Morrisa A, Norris KA. Colonization by *Pneumocystis jirovecii* and its role in disease. Clin Microbiol Rev. 2012;25:297–312.

Nannizzi A. Richerche sull'origine saprofita del funghi delle tigne. 2. Gymnoascus gypseum sp. n. forma ascofora del Sabouraudites (Achorion) gypseum. (Bodin) Ota et Langeron. Tai Accad Fisioer Siene. 1927;10:89–97.

Natarajan V, Pushkala S, Karuppiah VP, Prasad PV. Antidermatophytic activity of *Azardirachta indica* (Neem) by invitro study. Indian J Pathol Microbiol. 2002;45:111–3.

Negroni R. Clinical spectrum and treatment of classic histoplasmosis. Rev Iberoam Micol. 2000;17:159–67.

Negroni R. Historical aspects of dermatomycoses. Clin Dermatol. 2010;28:125–32.

Negroni R. Cryptococcosis. Clin Dermatol. 2012;30:599–609.

Nenoff P, Erhard M, Simon JC, Muylowa GK, Herrmann J, Rataj W, Gräser Y. MALDI-TOF mass spectrometry – a rapid method for the identification of dermatophyte species. Med Mycol. 2013;51:17–24.

Nosanchuk JD. Protective antibodies and endemic dimorphic fungi. Curr Mol Med. 2005;5:435–42.

Nosanchuk JD, Gacser A. *Histoplasma capsulatum* at the host-pathogen interface. Microbes Infect. 2008;10:973–7.

O'Gorman CM, Fuller HT, Dyer PS. Discovery of asexual cycle in the opportunistic fungal pathogen *Aspergillus fumigatus*. Nature. 2009;457:471–4.

O'Kinealy F. Localised psorospermosis of the mucous membrane of the septum nasi. Proc Laryng Soc London. 1903;10:109–12.

Oberoi JK, Wattal C, Aggarwal PK, Khanna S, Basu AK, Verma K. Pulmonary coccidiomycosis in New Delhi, India. Infection. 2012;40:699–702.

Onozaki M, Makimura K, Satoh K, Hasegawa A. Detection and identification of genotypes of *Prototheca zopfii* in clinical samples by quantitative PCR analysis. Jpn J Infect Dis. 2013;66:383–90.

Ophüls W. Observations on a pathogenic mould formerly described as a protazoan: (*Coccidioides immitis, Coccidioides pyogenes*). J Exp Med. 1905;6:443–86.

Orlowski M. Mucor dimorphism. Microbiol Rev. 1991;55:234–58.

Ota M, Langeron M. Nouvelle classification des Dermatophytes. Ann Parasitol. 1923;1:305–6.

Pal M. Isolation of Candida tropicalis from a case of empyema. Curr Sci. 1987;57:1135–6.

Pal M. Ringworm in a lion (*Panthera leo*) caused by *Microsporum gypseum*. Wildlife Health. 1988;6:1–4.

Pal M. Feline meningitis due to *Cryptococcus neoformans* var. *neoformans* and review of feline cryptococcosis. Mycoses. 1991;34:313–6.

Pal M. Guttural pouch mycosis in horse. The first reported case in India. Rev Iberoam Micol. 1996;13:31–2.

Pal M. Mycotic mastitis in buffalo (*Bubalus bubalis*) caused by *Candida tropicalis*. Buffalo J. 1997;13:91–4.

Pal M, Lee CW. Spontaneous pneumonia in a goat due to *Candida albicans*. Korean J Vet Clin Med. 1999;16:467–8.

Pal M, Lee CW. *Trichophyton verrucosum* infection in a camel and its handler. Korean J Vet Clin Med. 2000;17:293–4.

Pal M, Mehrotra BS. Cryptococcal mastitis in dairy animals. Mykosen. 1983a;26:615–6.

Pal M, Mehrotra BS. Occurrence and etiologic significance of *Cryptococcus neoformans* in a cutaneous lesion of a cat. Mykosen. 1983b;26:608–10.

Pal M, Dahiya SM, Lee CW. Family pets as a source of *Microsporum canis*. Korean J Vet Clin Med. 1990;7:151–5.

Pan S, Cole GT. Molecular and biochemical characterization of a *Coccidioides immitis*-specific antigen. Infect Immun. 1995;63:3994–4002.

Panja G. A case of generalized blastomycosis. Ind Med Gaz. 1925;60:475–6.

Papp T, Palaègyi Z, Ferenczy L, Vaègvoëlgyi C. The mitochondrial genome of *Mucor piriformis*. FEMS Microbiol Lett. 1999;171:67–72.

Pappagianis D, Zimmer BL. Serology of coccidioidomycosis. Clin Microbiol Rev. 1990;3:247–68.

Park SJ, Mehrad B. Innate immunity to *Aspergillus* species. Clin Microbiol Rev. 2009;22:535–51.

Pathak RC, Singh CM. Occurrence of *Candida sp.* in crops of fowls. Ind J Microbiol. 1962;2:89–90.

Peano A, Min AM, Beccati M, Menzano A, Pasquetti M, Gallo MG. Use of western blot to study *Microsporum canis* antigenic proteins in canine dermatophytosis. Mycoses. 2011;54:223–9.

Peng T, Orsborn KI, Orbach MJ, Galgiani JN. Proline-rich vaccine candidate antigen of *Coccidioides immitis:* conservation among isolates and differential expression with spherule maturation. J Infect Dis. 1999;179:518–21.

Platauf AP. Mycosis mucorina. Virchows Arch. 1885;02:543–64.

Posadas A. Um nuevo caso de micosis fungoidea com psorospermias. An Circ Med Argent. 1892;15:585–97.

Prabhu RM, Patel R. Mucormycosis and entomophthoramycosis: a review of the clinical manifestations, diagnosis and treatment. Clin Microbiol Infect. 2004;10:31–47.

Qiao J, Liu W, Li R. Antifungal resistance mechanisms of *Aspergillus*. Jpn J Med Micrbiol. 2008;49:157–63.

Radaeli F. Micosi del piede da Monosporium apiospermum. Sperimentale. 1911;52:109.

Randhawa HS, Chowdhary A, Kathuria S, Roy P, Misra DS, Jain S, Chugh TD. Blastomycosis in India: report of an imported case and current status. Med Mycol. 2013;51:185–92.

Ranganathan S, Arun Mozhi Balajee S, Mahendra Raja S. A survey of dermatophytosis in animals in Madras, India. Mycopathologia. 1997–1998;140:137–40.

Rao MAN. Rhinosporidiosis in bovines in Madras Presidency with a discussion in the probable modes of infection. Indian J Vet Sci. 1938;8:187–98.

Rao SR. Rhinosporidiosis in equines in Bombay state. Bombay Vet College Mag. 1951;2:1–4.

Rawal IJ, RaoBhan LN, Dutta PC. *B. dermatitidis* in tufted pochard Aythya fuligula. Ind Vet J. 1988;65:965–96.

Rivera ZS, Losada L, Nierman WC. Back to the future for dermatophyte genomics. MBio. 2012;3:e00381–12.

Rokas A, Gibbons JG, Zhou X, Beauvais A, Latg JP. The diverse applications of RNA-seq for functional genomic studies in *Aspergillus fumigatus*. Ann NY Acad Sci. 2012;1273:25–34.

Roncal T, Ugalde U. Conidiation induction in *Penicillium*. Res Microbiol. 2003;154:539–46.

Sabouraud R. Les teignes. Paris: Masson & Co; 1910.

Saccardo PA. Sylloge fungorum ommium hucusque cognitorum. Borntragen (Leipzig). 1911;4:282.

Saccente M, Woods GL. Clinical and laboratory update on blastomycosis. Clin Microbiol Rev. 2010;23:367–81.

Sadana JR, Kalra DS. A note on pulmonary phycomycosis in a pig. Indian J Anim Sci. 1973;43:961–4.

Sanfelice F. Contributo alla morfologia e biologia dei blastomiceti che si sviluppano nei succhi de alcuni frutti. Ann Ist Ig R Univ Roma. 1894;4:463–9.

Sargeant K, Sheridan A, O'Kelly J, Carnaghan RB. Toxicity associated with certain samples of groundnuts. Nature. 1961;192:1096–7.

Saubolle MA, McKellar PP, Sussland D. Epidemiologic, clinical, and diagnostic aspects of coccidioidomycosis. J Clin Microbiol. 2007;45:26–30.

Savio J, Srinivasa H, Mathew J, Chakraborty A, McNamara DR, Hemashettar BM, Padhye AA. Coccidioidomycosis in India: report of a third imported case. J Mycol Med. 2011;21:33–6.

Scaccabarozzi L, Locatelli C, Pisoni G, Manarolla G, Casula A, Bronzo V, Moroni P. Epidemiology and genotyping of *Candida rugosa* strains responsible for persistent intramammary infections in dairy cows. J Dairy Sci. 2011;94:4574–7.

Schenck BR. On refractory subcutaneous abscess caused by a fungus possibly related to the Sporotricha. Bull Johns Hopkins Hosp. 1898;9:286–90.

Scully C, El-Kabir M, Samaranayake LP. Candida and oral Candidosis: a review. Crit Rev Oral Biol Med. 1994;5:125–57.

Seeber GR. Un nuevo esporozuario parasito del hombre. Dos casos encontrados en polypos nasales. Thesis. Buenos Aires: Universidad Nacional de Buenos Aires, Imprenta Libreria Boullosa; 1900.

Segretain G. *Penicillium marneffei* n. sp., agent d'une mycose du systeme reticuloendothelial. Mycopathol Mycol Appl. 1959;11:327–53.

Sharma NL, Sharma RC, Gupta ML, Singh P, Gupta N. Sporotrichosis: study of 22 cases from Himachal Pradesh. Ind J Dermatol Venereol Leprol. 1990;56:296–8.

Shingu-Vazquez M, Traven A. Mitochondria and fungal pathogenesis: drug tolerance, virulence, and potential for antifungal therapy. Eukaryot Cell. 2011;10:1376–83.

Shubitz LF. Comparative aspects of coccidioidomycosis in animals and humans. Ann NY Acad Sci. 2007;1111:395–403.

Singh MP, Singh CM. *Trichophyton simii* infection in poultry. Vet Rec. 1972;90:218.

Skerlev M, Miklic P. The changing face of *Microsporum* sp. infections. Clin Dermatol. 2010;28:146–50.

Smith F. The pathology of bursattee. Vet J. 1884;19:16–7.

Spellberg BJ, Filler SG, Edwards JE. Current treatment strategies for disseminated Candidiasis. Clin Infect Dis. 2006;42:244–51.

Sriranganadane D, Waridel P, Salamin K, Feuermann M, Mignon B, Staib P, Neuhaus JM, Quadroni M, Monod M. Identification of novel secreted proteases during extracellular proteolysis by dermatophytes at acidic pH. Proteomics. 2011;11:4422–33.

Stenwig H, Taksdal T. Isolation of *Epidermophyton floccosum* from a dog in Norway. Sabouraudia. 1984;22:171–2.

Stringer JR, Cushion MT. The genome of *Pneumocystis carinii*. FEMS Immunol Med Microbiol. 1998;22:15–26.

Sutton DA. Diagnosis of coccidioidomycosis by culture, safety considerations, traditional methods, and susceptibility testing. Ann NY Acad Sci. 2007;1111:315–25.

Tambor JH, Guedes RF, Nobrega MP, Nobrega FG. The complete DNA sequence of the mitochondrial genome of the dermatophyte fungus *Epidermophyton floccosum*. Curr Genet. 2006;49:302–8.

Tchernev G, Penev PK, Nenoff P, Zisova LG, Cardoso JC, Taneva T, Ginter-Hanselmayer G, Ananiev J, Gulubova M, Hristova R, Nocheva D, Guarneri C, Kanazawa N. Onychomycosis: modern diagnostic and treatment approaches. Wien Med Wochenschr. 2013;163:1–12.

Terreni AA, Gregg Jr WB, Morris PR, DiSalvo AF. *Epidermophyton floccosum* infection in a dog from the United States. Sabouraudia. 1985;23:141–2.

Tewari RP. Studies on some aspects of mycoses. M.V.Sc. thesis. Submitted to Agra University, Agra; 1962.

Tewari RP. *Trichophyton simii* infection in chickens, dogs and man in India. Mycopath Mycol Appl. 1969;39:233–8.

Thakur SK, Verma BB. Dermatomycosis due to M. gypseum in goats. Indian Vet J. 1984;61:1083.

Thakur DK, Misra SK, Chaudhri PC. Prevalence of dermatophytosis in sheep in Northern India. Mykosen. 1983;26:271–2.

Theiler (1906) cited by Karunaratne WAE. Rhinosporidiosis in man. London: The Athlone press; 1964. p. 146.

Thompson DS, Carlisle PL, Kadosh D. Coevolution of morphology and virulence in *Candida* species. Eukaryot Cell. 2011;10:1173–82.

Trofa D, Gcser A, Nosanchuk JD. *Candida parapsilosis*, an emerging fungal pathogen. Clin Microbiol Rev. 2008;21:606–25.

Van de Sande WWJ, Fahal AH, Goodfellow M, Mahgoub ES, Welsh O, Zijlstra EE. Merits and pitfalls of currently used diagnostic tools in Mycetoma. PLoS Negl Trop Dis. 2014;8:e2918.

Vanek J, Jirovec O. 'Interstitieller' Plasmazellenpneumonie der Fruhgeborenen, verursacht durch *Pneumocystis carinii*. Zbl Bakt I Abt Orig. 1952;158:120.

Vanittanakom N, Cooper CR, Fisher MC, Sirisanthana T. *Penicillium marneffei* infection and recent advances in the epidemiology and molecular biology aspects. Clin Microbiol Rev. 2006;19:95–110.

Verghese S, Arjundas D, Krishnakumar KC, Padmaja P, Elizabeth D, Padhye AA, Warnock DW. Coccidioidomycosis in India: report of a second imported case. Med Mycol. 2002;40:307–9.

Vermout S, Baldo A, Tabart J, Losson B, Mignon B. Secreted dipeptidyl peptidases as potential virulence factors for *Microsporum canis*. FEMS Immunol Med Microbiol. 2008;54:299–308.

Vilela R, Mendoza L. The taxonomy and phylogenetics of the human and animal pathogen *Rhinosporidium seeberi*: a critical review. Rev Iberoam Micol. 2012;29:185–99.

Vollmer T, Störmer M, Kleesiek K, Dreier J. Evaluation of novel broad-range real-time PCR assay for rapid detection of human pathogenic fungi in various clinical specimens. J Clin Microbiol. 2008;46:1919–26.

Weitzman I, Summerbell RC. The dermatophytes. Clin Microbiol Rev. 1995;8:240–59.

West GS. Algae, vol. 1. Cambridge: Cambridge University Press; 1916. p. 475.

Wetzel J, Burmester A, Kolbe M, Wöstemeyer J. The mating-related loci sexM and sexP of the zygomycetous fungus *Mucor mucedo* and their transcriptional regulation by trisporoid pheromones. Microbiology. 2012;158:1016–23.

Wheat LJ. Approach to the diagnosis of the endemic mycoses. Clin Chest Med. 2009;30:379–89.

White CJ. Ringworm as it exists in Boston. J Cutan Genitourin Dis. 1899;17:1–17.

Willger SD, Grahl N, Cramer RA. *Aspergillus fumigatus* metabolism: clues to mechanisms of in vivo fungal growth and virulence. Med Mycol. 2009;47 Suppl 1: S72–9.

Witkamp J. Bijdrage tot de kennis van de *Hyphomycosis destruens*. Ned Ind Blad voor Diergeneeskd en Dierenteelt. 1924;36:229–45.

Woods JP. Knocking on the right door and making a comfortable home: *Histoplasma capsulatum* intracellular pathogenesis. Curr Opin Microbiol. 2003;6:327–31.

Worek M, Kwiatkowska A, Ciesielska A, Jaworski A, Kaplan J, Miedziak B, Deregowska A, Lewinska A, Wnuk M. Identification of dermatophyte species using genomic in situ hybridization (GISH). J Microbiol Methods. 2014;100:32–41.

Yegneswaran PP, Sripathi H, Bairy I, Lonikar V, Rao R, Prabhu S. Zoonotic sporotrichosis of lymphocutaneous type in a man acquired from a domesticated feline source: report of a first case in southern Karnataka, India. Int J Dermatol. 2009;48:1198–200.

Yoon HJ, Clemons KV. Vaccines against *Coccidioides*. Korean J Intern Med. 2013;28:403–7.

Youssef D, Shams W, Ganote CE, Al-Abbadi MA. Negative image of *Blastomyces* on Diff-Quik Stain. Acta Cytol. 2011;55:377–81.

Yu J. Current understanding on aflatoxin biosynthesis and future perspective in reducing aflatoxin contamination. Toxins. 2012;4:1024–57.

Yunus AW, Razzazi-Fazeli E, Bohm J. Aflatoxin B1 in affecting broiler's performance, immunity, and gastrointestinal tract: a review of history and contemporary issues. Toxins. 2011;3:566–90.

Zopf and Kühn (1880) cited by Krüger W. Kurz charakteristik einiger niederer organismen in saftflusse der laubbäume. I. Über einen neuen pilztypus, repräsentiert durch die gattung Prototheca (Pr. Muriformis et P. zopfii). Hedwigia 1894;33:241–51.

Zschokke E. Ein Rhinosporidium beim pferd. Schweiz Arch Tierheilkd. 1913;55:641–50.

5 Collection and Transport of Clinical Material for Isolation of Fungi

The collection of proper clinical specimens with sterile instruments and their timely shipment with appropriate arrangements into the laboratory is a crucial matter for isolation and identification of fungi. The correct type of specimen with sufficient quantity is required for proper identification. Table 5.1 describes the types of clinical specimens that can be collected from different body systems and types of fungal infection that can be identified from those specimens. The collection methods of clinical materials and their despatch are mentioned below.

Cerebrospinal fluid (CSF): The CSF is collected slowly by lumbar puncture between third and fourth vertebrae with a needle (18–20 s.w.g.) and a stylet (10 cm). However, in animals, the collection of CSF is not always safe. The animals have increased CSF pressure and the collection of CSF rapidly decreases the pressure with fatal consequences.

Eye: The corneal swabs are collected by holding the palpebra apart and gently swabbing the surface with the help of sterile cotton swabs. The corneal scrapings are taken from the base and margin of corneal ulcers with a sterile spatula under local anaesthesia. The swabs and scrapings are put into Stuart transport medium or sterile Sabouraud broth for transport. The anterior chamber fluid of eye can be collected immediately after death of the animal. It is done by inserting a small gauge needle through the cornea into the anterior chamber and aspirating the fluid into a 3 or 5 ml syringe.

Ear and nose: The ear and nasal swabs are collected by gently mopping the ear or nasal canal with the help of sterile cotton swabs. The swabs should be moistened with transport medium before collection. The collected swabs are put into Stuart transport medium or sterile Sabouraud broth or BHI broth for transport. The fine-needle aspirates from nasal polyp can be collected with a long needle. In rhinosporidiosis, nasal scrape is preferred over the fine-needle aspirates because the lesions bleed easily.

Gastrointestinal tract: The faeces should be collected in sterile, wide-mouth and screw-capped short jars. Rubber caps should be avoided because the gas generated in the collected faeces may blow the cap. The faeces and urine should not be mixed together. Rectal or cloacal swabs can be collected as described above. Gastric lavages are collected by flushing the empty stomach with sterile water. The washings are taken into sterile screw-capped glass containers.

Respiratory system: In human the sputum is collected in empty stomach in a screw-capped wide-mouth sterile container. In full stomach the sample may be contaminated with bacteria and other saprophytic fungi. The fresh single cough specimen expectorated in the early morning is the most ideal for fungal investigation. In animals, the sputum is collected with small throat swabs as described above. The sputum should be transported and processed rapidly to reduce the contamination level. The bronchoalveolar lavage (BAL) is preferably collected with a bronchoscope from human patients who cannot produce

I. Samanta, *Veterinary Mycology*, DOI 10.1007/978-81-322-2280-4_5,

Table 5.1 Types of clinical samples collected for fungal identification

Body parts/ system	Clinical specimens	Fungal infection
Central nervous system	Cerebrospinal fluid	*Cryptococcus, Coccidioides, Histoplasma, Blastomyces, Histoplasma capsulatum* var. *capsulatum, Penicillium marneffei, Sporothrix*
Eye	Swabs, corneal scrapings, anterior chamber fluid	*Blastomyces*
Ear	Swab	*Aspergillus*
Nose	Swabs, exudates, nasal scrape, fine-needle aspirates biopsy from nasal polyp	*Rhinosporidium, Aspergillus fumigatus* (horses), *Cryptococcus*
Oral cavity	Swabs, oropharyngeal washing, palatal papule scrapings	*Penicillium marneffei*
Gastrointestinal tract	Faeces, gastric lavage	*Histoplasma capsulatum* var. *capsulatum, Penicillium marneffei, Rhizopus*
Respiratory system	Sputum, Bronchoalveolar lavage, tracheal aspirate	*Aspergillus, Blastomyces, Coccidioides, Histoplasma capsulatum* var. *capsulatum, Mucor, Penicillium marneffei, Rhizopus, Sporothrix*
Urinary tract	Urine	*Blastomyces, Cryptococcus, Histoplasma capsulatum* var. *capsulatum, Penicillium marneffei, Sporothrix*
	Blood	*Candida, Cryptococcus, Histoplasma capsulatum* var. *capsulatum, Mucor, Penicillium marneffei, Rhizopus, Sporothrix*
Udder/ mammary gland	Milk	*Aspergillus, Candida, Cryptococcus, Mucor*
Aborted foetus	Abomasal content, liver, lung	*Aspergillus, Rhizopus*
Vagina	Swabs/discharge	*Rhizopus*
Lymph node	Aspirates, biopsy	*Blastomyces, Coccidioides, Cryptococcus, Penicillium marneffei, Pythium insidiosum*
Joint	Synovial fluid	*Blastomyces, Sporothrix*
Nail	Clippings	*Epidermophyton, Trichophyton, Mucor*
Skin	Scrapings, crust	*Trichophyton, Microsporum, Mucor, Aspergillus* (horse), *Candida, Coccidioides, Cryptococcus, Penicillium marneffei, Pythium insidiosum, Rhizopus*
	Pus/exudates/fine-needle aspirates from lesion/nodule	*Blastomyces, Coccidioides, Histoplasma capsulatum* var. *farciminosum, Madurella mycetomi, Sporothrix*
	Hair	*Microsporum, Trichophyton*
Viscera	Tissues	*Blastomyces, Candida, Cryptococcus, Histoplasma capsulatum* var. *capsulatum, Madurella mycetomi, Penicillium marneffei, Pythium insidiosum, Rhinosporidium, Rhizopus, Sporothrix*

sputum. However, flexible fiberoptic bronchoscopy is not a sterile procedure. Contamination can take place from upper airways and some microbes colonise in the tracheobronchial tree which is also detected in BAL samples.

Urinary tract: 'Mid-stream' urine collected during early morning is the best specimen for mycological investigation and should be sent to the laboratory immediately. It is a good media for growth of microbes. So chances of bacterial contamination are maximum.

Blood: The blood sample should be collected from animals by venipuncture. In large animals jugular vein or caudal vein is preferred. In pigs and birds, vena cava and wing vein (brachial vein) are selected, respectively. The skin of the collection site should be shaved and cleaned with 70 % ethanol and dried before collection. Whole blood (with anticoagulant) in triplicate is used for isolation of fungi. The serum is used for serological diagnosis of fungi and testing with paired sera (sera collected during acute phase and convalescent phase) can confirm the infection. After collection the serum is preserved with merthiolate (1:1,000) or carbolic acid (0.5 %).

Mammary gland: Milk from the mastitic animals is collected for fungal investigations. The udder is washed with antiseptic solution and the teats are cleaned with ethanol (70 %) swabs before collection. The primary stripping is discarded and later 10–15 ml of mid-stream milk is collected in a sterile and capped container.

Aborted foetus: The abomasal content is collected in sterile, wide-mouth glass bottle and the organs are taken in sterile petri dishes.

Vagina: The area should be disinfected with antiseptic and dried before collection of the vaginal discharge. The discharge is collected in a sterile screw-capped glass container with a pipette.

Joint: The synovial fluid in sufficient amount is collected from the joint with sterile needle and syringe and is transported in a screw-capped bottle.

Nail: The nails are swabbed with 70 % ethanol and dried before collection. The affected nail bed is exposed by removing the onycholytic nail plate with a nail clipper. They are wrapped in brown paper (or coloured paper) and are kept in a tight container preferably without moisture for transport into the laboratory. The powdery nail shavings are sent in glass plates.

Skin: The lesion area should be cleaned with 70 % ethanol to remove the bacterial contamination. The scales from the active lesions should be collected with a blunt scalpel (or toothbrush). The scrapings are packed in envelope or brown paper and put into airtight container for transport. To collect the exudates or pus from the skin lesion, the moist cotton swabs are gently rubbed over the lesion surface. After collection the swabs are placed into glass tubes with Sabouraud broth. The pus can be also collected by fine-needle aspiration technique. After collection of pus or exudate in sterile syringe, the needle should be discarded and the syringe should be capped before transport to reduce the environmental contamination. The skin biopsies should be collected both from periphery and centre of the lesion. The punch biopsy technique can be used under local anaesthesia.

Hair: The affected hairs should be plucked from the lesion without breakage. The use of Mackenzie's hair brush, a hard-bristled hand brush (2.5 in. × 1.5 in.), produces better isolation of dermatophytes than the hair plucking. In human, tweezer (forcep) is used to pluck the hairs. About 10–15 hairs are collected and placed into brown paper or sealed envelope for transport. The hair soiled with pus or exudates should not be collected.

5.1 Transport of Clinical Materials

All the clinical materials should be packed properly according to the guidelines of dangerous goods regulations (DGR, available in http://www.iata.org/publications/dgr). The packaging should be strong enough so that no sample is leaked from the container. Three-tier packaging which consists of a primary container (leak proof), secondary packaging and a rigid outer packaging is recommended. The dry ice or pre-frozen pack can be used surrounding the secondary packaging. The fungal agents causing animal or human diseases are not included in the DGR list of infectious agents affecting animals (UN 2900) or human (UN 2814) except *Coccidioides immitis* (culture only). So the packaged clinical materials can be labelled as 'Diagnostic specimen'/'Clinical specimen'/'Biological substance category B (UN 3373)' which should be clearly visible.

6 Diagnostic Techniques for Fungi

The conventional and modern diagnostic techniques which are used for identification of fungi are discussed below.

(a) Direct examination: It provides rapid information about the presence or absence of fungi. Accordingly the treatment of the affected animal or human may be started. The direct examination depends on the quality of the sample.

   (i) Smears are prepared from the clinical samples such as sputum, blood, exudates, urine deposit, tissue impression, etc. in triplicate. The first two smears are stained with Giemsa, Gram's or Wright's stain (*Penicillium*). The third smear is kept as reserve. Gram stain effectively stains the yeast cells (*Candida, Histoplasma* yeast phase). The yeast cells are also demonstrated by the negative staining with Diff-Quik (DQ) which demonstrates the unstained yeast cells present in the stained background. The fine-needle aspirates can be stained with the May –Grunwald–Giemsa method.

   (ii) Wet mount: The clinical materials such as skin scrapings, hair including the hair bulb, nail fragments, sputum and pus can be observed under microscope by wet mount with 10 % KOH preparation. The wet mount preserves fungal morphology without any distortion and helps in identification but lacks sensitivity. The KOH acts as a clearing agent which digests the proteinaceous host cell debris and leaves the sugar-containing fungal cell wall. The digestion of keratin with the KOH can be speed up with heat. An alternative preparation is 20 % KOH with 36 % DMSO that provides rapid diagnosis without heating and the specimens can be preserved for a long time. Addition of Parker ink with 10 % KOH (1:10) or chlorazol azole black (0.1 %) further improves the visualisation. Currently Parker ink is not available and chlorazol is mostly preferred. Chlorazol helps in identification of hyphae by staining the carbohydrate-rich wall without staining the contaminants such as cotton or elastic fibres.

   (iii) Fluorescent staining: Treatment of specimens with calcofluor white or blankophor or acridinium orange can also produce early diagnosis with better sensitivity than wet mount. These fluorochrome stains bind with chitin of fungal cell wall and produce fluorescence under ultraviolet light. The arthrospores appear as brightening structures. However, it can also non-specifically stain the plant and fatty substances (Fig. 6.1).

   (iv) Capsule staining: Capsule of *Cryptococcus* can be demonstrated by negative staining with India ink, nigrosin and Romanowski (Fig. 6.2). Romanowski stain produces clearer capsule against

I. Samanta, *Veterinary Mycology*, DOI 10.1007/978-81-322-2280-4_6,

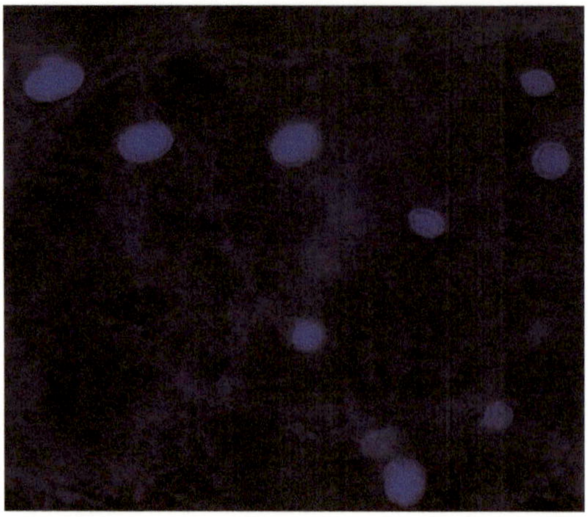

Fig. 6.1 *Prototheca zapfiri* in calcofluor white staining (1,000×) (Photograph courtesy: Prof. Cornelia Lass-Flörl, Medizinische Universität Innsbruck, Austria)

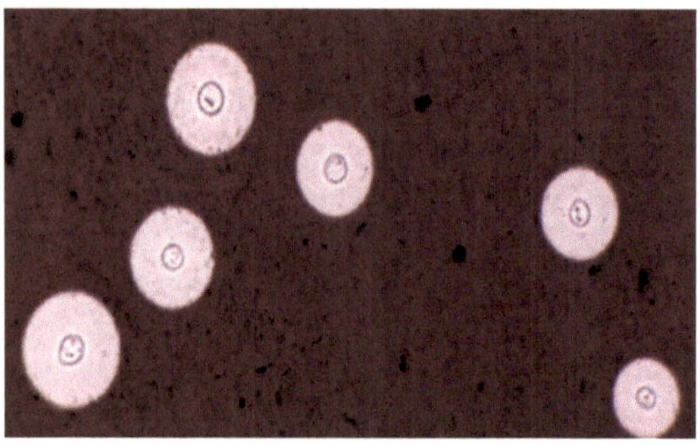

Fig. 6.2 Encapsulated *Cryptococcus* in India ink preparation (400×) (Photograph courtesy: Prof. P. P. Gupta, Ex-Director Veterinary Research, Punjab Agricultural University, Punjab)

the lightly stained background. The capsule itself can be stained with Mayer's mucicarmine stain.

(b) Slide culture technique: This technique helps in detection of detailed fungal morphology such as arrangement of hyphae, pattern of conidia formation, etc. A clean glass slide is placed over a V-shaped, sterile glass rod in a petri dish. A piece of agar block (plug/1 cm^2) is cut from the already prepared Sabouraud dextrose agar plate. The agar plug is placed over the slide. The fungus to be studied is inoculated at the four corners of the plug and a sterile coverslip is placed over the plug. A filter paper soaked with sterile distilled water is placed in the plate to prevent the drying of the agar. The lid is closed and the plate is kept at 25 °C for 5–7 days. When the sporulation occurs, the coverslip is taken out carefully and placed over a drop of mounting fluid (Lactophenol cotton blue/Narayan stain) in a clean glass slide keeping the fungal growth side downwards. Alternatively the mounting fluid is added over the used slide and a clean coverslip is placed over it. The slide is observed under low and high dry magnification.

(c) Isolation of pathogenic fungi: Isolation of fungi is still considered as gold standard method for diagnosis in spite of the fact that it is time consuming and it requires expertise. The conventional solid medium includes Sabouraud dextrose agar with cycloheximide (actidione, 400 mg/L) and chloramphenicol (50 mg/L). The cycloheximide inhibits the growth of bacteria, yeasts and certain moulds. So it is not incorporated in the media intended for isolation of *Aspergillus, Cryptococcus*, etc. The incubation time and temperature varies with the fungal species. However, the complete mould growth requires 2–5 weeks time. The yeasts are isolated in blood agar, corn meal agar, Sabouraud's dextrose agar (with antibiotics), honey agar, brain–heart infusion agar and malt agar. The plates are incubated at 28–37 °C for 2 days–2 weeks. Dimorphic conversion (mycelium to yeast phase) can take place when the subculture of mycelium can be performed in brain–heart infusion agar with blood at 35–37 °C. Identification of fungi depends on front and reverse pigmentation of the colonies, characteristics of the colonies and their microscopic appearance (Fig. 6.3).

(d) Germ tube test: The test is performed to confirm the presence of *Candida albicans* or *C. tropicalis*. A single candidal colony is

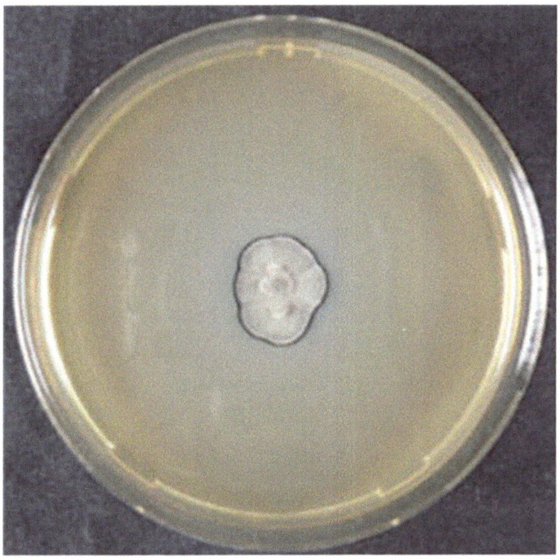

Fig. 6.3 Fungal colony (Photograph courtesy: Prof. Sybren de Hoog, CBS Fungal Biodiversity Centre, The Netherlands)

picked up and mixed with 0.5 ml of rabbit or sheep or bovine or human serum and incubated at 37 °C for 1–2 h. After the incubation, a drop of preparation is observed under the phase contrast or dry objective of the microscope. In positive case, small tubes projecting from the yeast cells without any constriction at the point of attachment are observed. This is a diagnostic feature of *C. albicans* and *C. tropicalis* (prolonged incubation for 3 h).

(e) Hair bait test: The test is performed to detect keratinophilic dermatophytes from the soil. The collected surface soil is kept in sterile petri dish up to half of the depth. The top layer of the soil is made wet by sprinkling sterile distilled water. The sterile human or horse hairs (20–30) are spread over the surface. The petri dish is incubated at 30 °C for a week. After 7 days, the plate is opened to detect any white/brown growth surrounding the hair. If the growth is detected, the hairs are observed under the microscope with mounting fluid and are transferred into Sabouraud dextrose agar (SDA) plates. The SDA plates with hairs are incubated again at 30 °C for a week to confirm the dermatophyte.

(f) Hair perforation test: This test can differentiate between *Microsporum canis–M. equinum* and *Trichophyton mentagrophytes–T. rubrum*. The child hairs are collected which should be preferably free from dust, oil and dandruff. The hairs are cut into small pieces (1–2 cm) and are placed into a glass petri dish. The petri dish along with hair is sterilised by autoclaving. The plate is filled with sterile distilled water with yeast extract (10 %) and small pieces of test fungus culture. The positive control plates are also prepared in similar way with *M. canis* or *T. mentagrophytes* culture. All the plates are incubated at 25 °C and are examined weekly for 4 weeks. Individual hair is placed on the mounting fluid and is observed under microscope to detect whether any hair perforation takes place like the positive control. If there is no hair perforation up to 30 days, it is considered as negative.

(g) Wood's lamp examination: The ectothrix dermatophyte infection of hair can be detected by ultraviolet rays (Wood's lamp). The ectothrix spores (majority of *Microsporum*) attached with the hair shaft produce a greenish-yellow fluorescence when exposed to Wood's lamp in a dark room. However, majority of *Trichophyton* species (except *T. schoenleinii*) are negative under Wood's lamp. This test is useful in long-haired animals where the dermatophyte lesions are not prominent in the skin. If the hairs are contaminated with oil-based lotions used for treatment, it may produce false-positive fluorescence under Wood's lamp.

(h) Animal inoculation technique: Animal inoculation technique is performed to isolate the fungi from clinical materials and to assess the pathogenicity of the fungal isolates subject to permission of animal ethics committee. Isolation of fungi in laboratory animals helps to obtain pure culture in artificial media. Table 6.1 describes the species of laboratory animals used for isolation of different pathogenic fungi. White mice and guinea pigs (for dermatophytes) are preferred for pathogenicity testing of the fungi. The animals are

Table 6.1 Use of laboratory animals for isolation of fungi

Laboratory animal	Fungi
Mice	*Aspergillus*, *Blastomyces*, *Coccidioides*, *Histoplasma*, *Sporothrix*, *Candida*
Guinea pigs	*Blastomyces*, *Coccidioides*, *Cryptococcus*, Dermatophytes, *Histoplasma*
Hamsters	*Blastomyces*
Rabbit	*Aspergillus*, *Coccidioides*, *Candida*
Rat	*Coccidioides*, *Sporothrix*

inoculated by intraperitoneal, intravenous, intratesticular and intracerebral routes. Intratesticular and intracerebral routes are used for pathogenicity assessment of *Sporothrix* and *Cryptococcus neoformans*, respectively.

(i) Antigen detection test: The antigen detection tests are useful in immunocompromised patients or in early stage of the infection when detectable antibodies are not produced. Galactomannan (GM) is the predominant antigen released by *Aspergillus fumigatus* which can be detected by latex agglutination test and sandwich ELISA. However, the GM detection assay is not specific for *Aspergillus* as it is cross-reacting with other fungi such as *Penicillium*, *Fusarium*, *Alternaria* and *Histoplasma*. Similarly, detection of D glucan (BDG) can give a presumptive diagnosis of *Aspergillus*, *Candida*, *Fusarium*, *Pneumocystis*, etc. Currently *Blastomyces* adhesin (BAD1) is the target antigen for specific detection of *Blastomyces* which do not cross-react with *Histoplasma*.

(j) Serological tests: ELISA-based serological tests can detect the specific fungal antibodies. However, serological tests are not so much reliable. The false-positive result occurs due to cross-reaction, superficial colonisation and persistence of the infection, and false-negative result occurs in immunocompromised patients producing low or undetectable level of antibodies.

(k) Molecular biological techniques: Conventional methods used in diagnostic mycology usually are less specific and more time consuming. Further, a trained mycologist is needed to identify the fungi morphologically. So molecular biological techniques are better alternative approach.

(i) Polymerase chain reaction (PCR): PCR is the process by which in vitro amplification of target DNA takes place, making their millions of copies. The amplified DNA becomes visible under ultraviolet light. The process is highly sensitive, specific and requires less time and little expertise. However, liberation of DNA form fungal cell is challenging due to the presence of a highly rigid cell wall which often produces false-negative result. Conventional PCR is developed for detection of dermatophytes (*T. rubrum*, *T. interdigitale*, *M. gypseum*, *M. canis*, *T. tonsurans*, *T. violaceum*, *E. floccosum*) using nuclear ribosomal DNA (rDNA), mitochondrial DNA (mtDNA), chitin synthase-1 (chs-1), topoisomerase II (TOP2) genes, small (18S rRNA) and large-subunit (28S rRNA) of ribosomal RNA as major target genes. Identification of *Candida* is possible by PCR using rRNA (5.8S, 18S, 28S), internal transcribed spacer (ITS) and intergenic spacer (IGS) region genes as target.

(ii) Molecular characterisation: PCR may produce false-positive result when conserved rRNA or other genes are targeted and airborne fungal contamination occurs. So, further confirmation of PCR amplicon is required by molecular characterisation. Randomly amplified polymorphic DNA (RAPD), PCR fingerprinting, amplified fragment length polymorphism (AFLP) and nucleotide sequencing have been successfully applied to species identification in dermatophytes but were mostly unable to discriminate between the strains. The sequencing of the internal transcribed spacer (ITS) and non-transcribed spacer

(NTS) region of the rRNA genes are used for phylogenetic analysis and identification of dermatophyte strains. Confirmation of *Aspergillus fumigatus* isolates by PCR-RFLP can be performed using *BccI*, *MspI* and *Sau3AI* restriction enzymes.

(iii) PCR-ELISA: Recent progress includes the use of PCR-ELISA which can specifically identify the PCR amplified product with the help of enzyme labelled probe producing colour reaction in positive cases. The uniplex-PCR-ELISA can identify *Trichophyton rubrum*, *T. interdigitale*, *T. tonsurans* and *T. violaceum* individually.

(iv) Real-time PCR: Real-time PCR is developed using Taqman probe or light cycler system for identification of *Candida, Coccidioides, Histoplasma capsulatum* var. *capsulatum*, *Mucor* and *Penicillium marneffei*.

(v) DNA microarray: DNA microarray is developed for simultaneous detection of several pathogenic fungi such as *Mucor*, *Aspergillus*, *Candida*, *Cryptococcus*, dermatophytes, etc.

(vi) Loop-mediated isothermal amplification (LAMP): It is developed for the confirmation of mycetoma caused by *Pseudallescheria boydii* which do not require any sophisticated instrument.

(l) Matrix-Assisted Laser Desorption Ionization Time-of-Flight Mass Spectrometry (MALDI-TOF-MS): MALDI-TOF-MS is used in identification of bacteria and fungi, and it has revolutionised the diagnostic microbiology. The method is specific, sensitive and quick to process, and it has relatively low cost than the nucleic acid sequencing. The technique can identify the fungi both from culture grown in SDA, blood agar and chrom agar and from the clinical specimens such as urine. The intact cells of the target fungi are directly streaked onto the MALDI target plate and overlaid with a matrix solution lysing the cells. The cell components after lysis are embedded into the matrix. Matrices act as ionising agent and it also helps in energy transfer from the laser to the analyte (cell components). The laser fire desorbs the analytes from the target plate. The resulting ions are accelerated in an electric field and focused to fly along the flight tube attached with a detector. Small ions move faster than the larger ones to reach the detector. The differences in the time required for the ions to reach the detector show differences in analysed peak (spectrum) which is specific for a species and subspecies. The species is identified by comparing the test spectra with references available in central database. The MALDI-TOF-MS is successfully used in differentiation between *Trichophyton* species of *Arthroderma benhamiae* and *Microsporum canis*, differentiation of *Candida* species, identification of different phenotypic variations of *Aspergillus* species and differentiation of filamentous fungal species including *Rhizopus*.

Appendix

Composition of Commonly Used Mounting Fluids/Stains

Andre and Hoyer's fluid (mounting fluid)

Arabic gum	15.0 g
Chloral hydrate	100.0 g
Glycerol	10.0 g
Distilled water	25.0 ml

Use: Primary study of fungal morphology

Giemsa stain

Giemsa	1.0 g
Glycerol	66.0 ml
Absolute methyl alcohol	66.0 ml

Commercially the above composition is available which is mixed with supplied buffer (1:9)
Use: Demonstration of yeast cells and moulds

Glycerine jelly (mounting fluid)

Gelatin	1.0 g
Glycerol	7.0 g
Distilled water (with 1 % phenol)	6.0 ml

Use: Primary study of fungal morphology

India ink

India ink	1.0 ml
Formalin (40 %)	9.0 ml

Use: Demonstration of capsule in *Cryptococcus*

KOH solution (10 %)

Potassium hydroxide (KOH)	10.0 g
Distilled water	100 ml

Use: Demonstration of dermatophytes from skin scrapings, hair including the hair bulb, nail fragments, sputum and pus

Lactophenol cotton blue (mounting fluid)

Phenol	20.0 g
Lactic acid	20 ml
Glycerol	40 ml
Distilled water	20 ml

After dissolving by gentle heat add 0.05 g cotton blue
Use: It is used for primary study of fungal morphology isolated from animals and human

Narayan stain (mounting fluid)

Methylene blue	0.5 ml
Dimethyl sulphoxide	7.0 ml
Glycerine	4.0 ml

Use: It is used for primary study of fungi and algae morphology isolated from animals and human

New methylene blue stain

Potassium oxalate	1.6 g
New methylene blue powder	0.5 g
Distilled water	100 ml

Use: Differentiation of viable and dead yeast cells. Viable cells will be unstained and dead cells will be stained as blue colour

I. Samanta, *Veterinary Mycology*, DOI 10.1007/978-81-322-2280-4,

Nigrosin stain

Nigrosin (granular)	10.0 g
Formalin (10 %)	100.0 ml

The solution is placed in hot water bath for 30 min and is filtered twice through double filter paper (Whatman No. 1)
Use: Demonstration of capsule in *Cryptococcus*

Phol stain

Formalin (4 %)	5.0 ml
Methylene blue (3 %)	0.3 ml
Glycerol	3.0 ml

Use: Morphology study of dermatophytes, yeasts and algae

Wright's stain

Wright's stain powder	0.3 g
Glycerol	3.0 ml
Methyl alcohol	97.0 ml

Use: Demonstration of yeast cells and moulds

Composition of Commonly Used Media in Diagnostic Mycology

Blood agar

Peptone	15.0 g
Liver extract	2.5 g
Yeast extract	5.0 g
Sodium chloride	5.0 g
Agar	15.0 g
Distilled water	1,000 ml
pH (at 25 °C)	7.4 ± 0.2

Use: Isolation of yeast

Corn meal agar

Corn meal	2.0 g
Tween 80 (1 %)	10 ml
Agar	15.0 g
Distilled water	1,000 ml

Use: Detection of chlamydospore-producing *Candida* and isolation of fungi

Czapek's Dox Agar

Potassium chloride (KCl)	0.5 g
Sodium nitrate ($NaNO_3$)	3.0 g
Di-potassium hydrogen phosphate (K_2HPO_4)	1.0 g
Magnesium sulphate ($MgSO_4$, 7 H_2O)	0.5 g
Ferrous sulphate ($FeSO_4$, 7 H_2O)	0.01 g
Sucrose/glucose	30.0 g
Agar	15.0 g
Distilled water	1,000 ml

Use: Isolation of yeast and mould

Dermatophyte test medium

Papaic digest of soybean meal	10.0 g
Glucose	10.0 g
Phenol red	0.2 g
Agar	20.0 g
Distilled water	1,000 ml
pH (at 25 °C)	5.5 ± 0.2

Use: Selective isolation of dermatophytes

Mycosel agar

Papaic digest of soybean meal	10.0 g
Dextrose	10.0 g
Agar	15.5 g
Cycloheximide	0.4 g
Chloramphenicol	0.05 g
Distilled water	1,000 ml

Use: Isolation of pathogenic fungi from contaminated clinical materials

Potato dextrose agar

Potatoes (peeled)	200.0 g
Dextrose	20.0 g
Agar	15.0 g
Distilled water	1,000 ml
pH (at 25 °C)	5.6 ± 0.2

Use: Isolation of fungi from clinical materials

Sabouraud dextrose agar (Emmon's)

Dextrose	20.0 g
Peptone	10.0 g
Agar	20.0 g
Distilled water	1,000 ml
pH	7.0 ± 0.2

Use: Maintenance of fungal culture

Sabouraud dextrose agar (Emmon's) with antibiotic

Dextrose	20.0 g
Peptone	10.0 g
Agar	20.0 g
Chloramphenicol (dissolved in 95 % ethanol)	50.0 mg
Distilled water	1,000 ml

Use: Isolation of fungi from clinical materials

Sabouraud dextrose agar (Emmon's) with antibiotic and cycloheximide

Dextrose	20.0 g
Peptone	10.0 g
Agar	20.0 g
Chloramphenicol	50.0 mg
Actidione	400.0 mg
Distilled water	1,000 ml

Use: Isolation of dermatophytes and dimorphic fungi from clinical materials

Sunflower seed agar (Pal's medium)

Pulverised sunflower seed	45.0 g
Chloramphenicol	100 mg
Agar	20.0 g
Distilled water	1,000 ml

Use: Isolation of *Cryptococcus neoformans*

Water agar

Agar	20.0 g
Distilled water	1,000 ml

Use: Cultivation and enumeration of some fungi

Glossary

Acropetal A chain of conidia in which the youngest conidium is at the tip and the oldest is at the base

Adiaconidia A large, globose, thick-walled conidium, usually produced by *Emmonsia parvum* in the lungs of human and animals

Aerial mycelium Hyphal elements growing above the agar surface

Aleuroconidium (plural, aleuroconidia) A conidium that develops as an expanded end of an undifferentiated hypha or on a short pedicel and is released by rupture of the supporting cell

Anamorph Asexual form of the fungus

Annellation Formation of ring-like structures at the conidiogenous end of a conidiophore

Annellide A conidiogenous cell that produces conidia in a basipetal way

Anthropophilic An organism that primarily infects human

Apophysis Funnel-shaped swelling of a sporangiophore which is present immediately below the columella

Appressorium A swelling on a germ tube or hypha which is used for attachment in an early stage of infection

Arthrospores Asexual spores produced by fragmentation of hyphae

Ascocarp A fruiting body containing asci and ascospores

Ascospore A haploid sexual spore formed in an ascus following meiosis

Ascus (plural, asci) A sac-like structure in which ascospores are produced. Asci are characteristic of Ascomycetes

Aspergilloma A fungus ball composed of *Aspergillus* hyphae and is located within the lung cavity

Autopsy Post-mortem examination of human or animal

Basidiospore A haploid spore produced on a basidium following karyogamy and meiosis

Basidium (plural, basidia) A cell that gives rise to a basidiospore

Basipetal The youngest conidium is found at the base of a chain and the oldest is found at the tip

Biopsy Collection of tissue from live animal or human

Bipolar budding Blastoconidia developing at the opposite poles of a parent cell

Blastospores Asexual spores produced by budding of yeast

Budding Asexual reproduction in which a bud is formed from the parent cell

Chlamydospore A conidium with a thickened cell wall that may be terminal or intercalary and serves the function of survival

Clavate Club shaped

Cleistothecium (plural, cleistothecia) An enclosed fruiting body that contains asci

Coenocytic (aseptate) hyphae Hyphae without septa

Collarette A remnant of a cell wall present at the tip of a phialide or around a sporangiophore

Columella A sterile, dome-shaped expansion at the apex of the sporangiophore

Conidiogenous cell A cell that produces conidia

Conidiophore A specialised hypha on which conidia are generated singly or in clusters

Conidium (plural, conidia) A nonmotile asexual spore which is produced from fungal mitosis. It is either unicellular (microconidium) or multicellular (macroconidium)

I. Samanta, *Veterinary Mycology*, DOI 10.1007/978-81-322-2280-4,

Dematiaceous Pigmented fungi

Denticle A small projection on which conidia are produced

Dichotomous Fungi having equal branching of hyphae, e.g. *Aspergillus*

Dimorphic Fungi having two different morphological forms, i.e. yeast and mould

Double septum A two-layered septum that undergoes centripetal separation (schizolysis) to release a conidium

Echinulate Covered with spines

Ectothrix Dermatophyte infection in which arthrospores are produced outside the hair shaft, e.g. *Trichophyton verrucosum*

Endothrix Dermatophyte infection in which arthrospores are produced inside the hair shaft, e.g. *Trichophyton schoenleinii*

Floccose Loose, cottony texture

Foot cell A basal cell of a conidiophore (*Aspergillus* and *Fusarium*)

Fungus (plural, fungi) Eukaryotic, heterogenous, unicellular to filamentous, spore bearing, chemoorganotrophic organisms which lack chlorophyll

Geniculate Bent like a knee

Geophilic Soil-inhabiting organisms

Germ tube Small tubes projecting from the yeast cells without any constriction at the point of attachment, e.g. *Candida albicans*

Gloiospora Conidia aggregated in slimy heads at the tip of an annellide or phialide

Gymnothecium (plural, gymnothecia) A fruiting body composed of loosely interwoven hyphae and asci

Habitat Natural environment of an organism for multiplication

Haustorium A special hyphal branch which is used for absorption of food

Heterothallic During sexual reproduction two fusion gametes are produced by different thallus

Hilum A scar at the base of a conidium

Holomorph Whole fungus which includes both the anamorph and the teleomorph state

Holothallic A mode of conidia formation when all the cell wall layers of the conidiogenous or sporogenous cells are involved

Homothallic During sexual reproduction two fusion gametes are produced by the same thallus

Hyaline Translucent or colourless

Hypha (plural, hyphae) Tube-like structure which makes the mycelium of a fungus

Intercalary Occurring within a hypha

Lanose Woolly

Merosporangium (plural, merosporangia) A small cylindrical sporangium with the sporangiospores aligned in a row

Metula (plural, Metulae) A sterile cell below phialides (e.g. *Aspergillus*, *Penicillium*)

Mould Mycelial fungus

Multipolar budding Blastoconidia developing at different sites on the surface of a parent cell

Muriform Conidia with longitudinal or transverse wall (e.g. *Alternaria*)

Mycelium (plural, mycelia) A mat of intertwined hyphae

Mycetism Infection of human or animals due to in vivo toxin production by the fungi

Mycosis (plural, mycoses) Fungal disease of human or animal

Mycotoxicosis Intoxication of human or animals due to consumption of mycotoxin-contaminated food or feed

Obclavate Club shaped in reverse

Ostiole An opening through which spores or conidia can escape

Pectinate Like the teeth of a comb

Pedicel A slender stalk

Penicillus (plural, penicillin) Brush-like conidiophore of *Penicillium*

Perfect stage Sexual state of the fungi

Phialide Sac-like structure that produces conidia (e.g. *Aspergillus*)

Phialoconidium A conidium produced from a phialide

Phragmoconidium A conidium having two or more transverse septa

Poroconidium A conidium produced through a small pore in a conidiogenous cell

Propagule A reproductive unit that gives rise to an organism

Pseudohyphae Sometimes the buds from the parent cell fail to detach and they produce a chain of elongated hyphae like filament called 'pseudohyphae'

Pseudomycelia Mass of pseudohyphae

Pyriform Pear shaped

Rhizoid Short branching hypha resembling a root (e.g. *Rhizopus*)

Saprobe (or saprotroph) An organism that obtains its nutrients by absorption of soluble organic compounds from nonliving or decaying organic matter, plant or animal

Sclerotium (plural, sclerotia) An organised mass of hyphae that remains dormant during unfavourable conditions (also called 'grain')

Septate Hyphae having septum

Septum (plural, septa) Cross wall in a mycelium or conidia

Sessile Direct hyphal attachment

Spherule Circular, thick-walled, closed structure containing endospores

Spinulose Covered with small spines

Spitzenkorper A structure present in fungal hyphae consists of small vesicles and it organises the fungal growth

Sporangiolum A small sporangium producing a small number of sporangiospores

Sporangium (plural, sporangia) Closed spherical structure containing sporangiospores

Sporodochium A cushion-shaped mass of hyphae bearing conidiophores

Sterigmata Specialised hypha bearing a conidium (e.g. *Aspergillus*)

Stolon A running hypha from which rhizoids and sporangiospores arise

Sympodial Conidiophores which produce conidia on geniculate or zig-zag basis (e.g. *Fonsecaea*)

Synnema (plural, synnemata) Erect macroscopic structure formed by fused conidiophores that bear conidia terminally, laterally or in both ways

Teleomorph A sexual state of fungi

Thallus (plural, thalli) Vegetative body of a fungus

Tuberculate Small spine-like projections/wart-like structures from a conidium

Uniserate Phialides arising directly from a vesicle (e.g. *Aspergillus*)

Verrucose Having many warts

Vesicle Swollen cell produced at the terminal end of conidiophores

Zoospores Motile asexual spores

Zygospores Thick-walled sexual spores

Index

I. Samanta, *Veterinary Mycology*, DOI 10.1007/978-81-322-2280-4,